Agents of Intimacy: Indian Aphrodisiacs, Psychedelics and Sex

Dr Anand Mahalingam

Published by Lulu.com (2022)

ISBN 978-1-4716-5252-3

Preliminary warnings and considerations

1. The legality of some of the substances mentioned in this book is complex, some substances being subject to varying restrictions in the jurisdictions of particular countries, while others have seen the recent lifting of restrictions regarding use. It needs to be emphasized that in this publication no one is being advised to consume any substance that is prohibited or illegal in the jurisdiction of the consumer.

2. Before consuming any substance, the consumer should obtain expert advice from a professional herbalist on dose-levels and possible negative effects arising from a combination with other substances.

3. People have a wide range of reactions to the effects of psychoactive substances. For some people, with particular dispositions, underlying conditions or genetic make-up, the effects of some compounds could be dangerous for mental or physical health.

4. Should any substance be consumed, the consumer is advised to first experiment with very small dose levels of **only** legal substances to ascertain whether or not the effect is desirable and to see if there is any negative reaction to the effect.

Preface

In 2017, through a mutual friend, I met Dr Anand Mahalingam, who worked throughout the latter part his life as a doctor; he is now retired. He lives in a small flat, not far from the sea, in a village on the coast in the state of Goa, India. He told me that over the course of a few years he had written a short book on aphrodisiacs. Over numerous cups of chai in his living room, stacked to the ceiling with books, we carefully went through his manuscript. Dr Mahalingam has travelled very widely during his life and has a wide range of interests: in philosophy, religion and history, and also in psychoactive plants. Dr Mahalingam has no interest in publicity and would like to remain out of public view. For this reason, he asked me if I could help facilitate the publication of this short book, which I was very happy to do.

Rajiv Pratilipi (editor), February 2022

CONTENTS **PAGE**

1. *Introduction*

This is a slightly odd book, being a compilation of disparate observations and lists of plants, which comprise five separate, though related, parts. It could be seen as a collection of notes and observations rather than as a comprehensive treatise. Each section of the book may be read independently. The main content of this publication consists of botanical identifications of Indian, multi-plant, aphrodisiac formulas, though some general observations on psychedelics and aphrodisiacs are also provided.

Part 1 (Sections 1–6) is largely anecdotal, providing some notes my own experiences with various—mostly psychedelic—substances. In this part there are also personal opinions and observations on the consumption of substances generally.

Part 2 (Sections 7–9) contains a brief synopsis of several well-known aphrodisiac plants and other aphrodisiac substances. There is a list of several chemicals and plants that I have not tried (Section 8.2). They were included, out of interest, as they feature in other publications on aphrodisiacs.

Part 3 (Sections 10–12) is a quite detailed survey of Indian, aphrodisiac plant formulas, perhaps of interest to botanists. A comparison between ancient and modern Āyurvedic, aphrodisiac formulas is provided (Section 11). Section 12 is the largest part of this book, included to inform the interested reader about the ingredients of around thirty, currently available aphrodisiac formulas, all of which I have taken, some just once, some on numerous occasions. As this is largely is a personal account, there is, unfortunately, only little information on aphrodisiacs for women in this book.

Part 4 (Section 13) contains information on a few non-India aphrodisiac formulas.

Part 5 (Section 14) contains descriptions of a few plants with monoamine oxidase inhibitors (MAOIs); one of the effects of MAOIs can be aphrodisiac.

Even though this book is concerned with both psychedelics and aphrodisiacs, the reader will notice that there is far more information on aphrodisiacs than on psychedelics. The two topics are paired as I believe there is an important overlap between these two broad classes of substances; I explain in Part 1 what I believe is the connection between them. I am neither an expert in botany or chemistry, so apologies to readers in advance for any errors in this text in that regard.

There are numerous publications on aphrodisiacs. A few of them, which are listed in the bibliography, were consulted in the course of writing this account.[1] However, most of the information in this monograph is based primarily on my own experiences with various substances, rather than being a catalogue of all known aphrodisiac substances. The survey in this book is not at all a complete survey of the topic.

---

[1] Choueke (1998); Lee and Lee (1994); Miller (1985); Rätsch and Müller Ebeling (2013); Selden (1979); Taberner (1985). Perhaps the most comprehensive catalogue is the excellent volume by Rätsch and Müller Ebeling (2013). Abel (1985) provides a detailed survey of the aphrodisiac (or otherwise) effects of a number of commonly used substances/drugs: alcohol, amphetamines, amyl nitrite, antidepressants (tricyclics and monoamine oxidase inhibitors [MAOIs]), antipsychotics (1. derivatives of phenothiazine [e.g. Thorazine], 2. butyrophenones [e.g. Haldol], 3. rauwolfia alkaloides [reserpine, as in *Rauwolfia serpentina*]), barbiturates, benzodiazepines (e.g. Valium, Librium), cocaine, coffee, hallucinogens, marihuana, methaqalone, narcotics, tobacco.

Most of the aphrodisiac plants and formulas included in this monograph are Indian in origin, as these are the ones I have mainly experimented with. There are, of course, numerous other aphrodisiac plants and formulas used by people in other parts of Asia, Europe, Africa and the Americas[2] that are not reported on in this short study, as they are largely outside my experience. One of the main purposes of this account is to provide the full details of the ingredients of Indian aphrodisiac formulas, which I have not seen published before.

Dr Anand Mahalingam, *magh pūrṇimā* (full-moon, 16th February), 2022

---

[2] Before the global regulation of plants and medicines became effective in the early 20th century, there was considerable interest in aphrodisiac formulas. See, for example, the advertisements for various aphrodisiac tinctures in the USA (Taberner 1985:76–83).

Part 1
Observations and Experiences with Psychedelics

2. *Inebriation in general*

I do not usually find accounts of people's experiences with psychoactive agents particularly interesting, but I include the biographical notes below to contextualise my observations and reports on a range of substances. I am sure some readers will have had far broader experience with particular psychedelic substances than I have had. The points made in the general discussion below of inebriation are not novel; I merely reiterate various observations that have been made previously by several commentators, which corroborate my own conclusions, arrived at during the course of the last half century.

Over a twelve-year period, between 2005 and 2017, I enjoyed a period of experimenting with a wide variety of primarily Āyurvedic aphrodisiac formulas, which are readily available on the internet or over the counter in pharmaceutical shops in India. After trying a couple of formulas that I found effective, I became curious to experiment with others. The report presented below is far from comprehensive in terms of the aphrodisiac formulas surveyed.

It should be noted that the effects of any substance or formula taken as an aphrodisiac may vary widely, even on the same consumer. Depending on several important factors—including set and setting, one's current hormonal condition, and other foods, beverages or inebriants recently consumed—the same substance or aphrodisiac formula may have a pronounced effect on one occasion, while on another may have little or no effect at all. The effect of any aphrodisiac is not guaranteed.

In the last five years or so, since the period of experimentation with aphrodisiac formulas, I have ceased to take them, having come to the conclusion that sustained use of aphrodisiacs is not particularly good for health. I have been able to detect slight, temporary, adverse effects on either my heart, liver or prostate gland from some formulas, but I remain generally uncertain as to which plant is having the negative effect in any particular formula. Though this is not to say that the journey was not interesting.

Concerning inebriation in general, it might be asked why we need to ingest inebriating substances at all. Why not just stay always sober? Is 'drug taking' not a symptom of something 'wrong' or 'amiss' in someone's life? The work of Ronald Siegel (2005) deserves due consideration in this regard: the evidence indicates that inebriation is an entirely normal activity, not only for humans in all nearly all cultures both now and historically, but also for many animals, birds and insects (see also Samorini 2002).

After decades of the study of numerous creatures, Siegel (2005:208) maintains that intoxication should be considered as a basic, instinctive drive, which he categorises as the fourth drive in humans, alongside the drives for sex, food and drink, which cannot be suppressed, as they are biologically inevitable. The pertinent issue with intoxication is how to become intoxicated safely, ensuring that setting, dose-levels, the individual's psychology and drug habits, and potential, negative cross-interference with other foods and medications or drugs are taken into adequate consideration.

Most people take drugs very regularly, whether that be tea, coffee, *yerba maté* (*Ilex paraguariensis*), *pān* (betel nut, *Areca catechu*), *khat* (*Chata edulis*), alcohol, tobacco or something else. Some people take substances for strong effects. It has been suggested by some commentators that perhaps 5%–10% of human populations—which is still a very large number of people—are practically 'born' with social problems. (I am aware that this figure or notion could be challenged.) Maybe they were born upside-down; who knows. Almost everything done to help them has little effect. Sometimes their condition is readily apparent in primary school, where they believed to have special needs. They may remain socially dysfunctional; they may have problems with personal reliability and working with other people. Such people are prone to taking drugs in excess. Their consumption of drugs may render them unfit to work. Such behaviour could be seen as a form of self-harm. Their anti-social behaviour requires the intervention of family, friends or health-care professionals. They draw attention to themselves through anti-social behaviour.

Some people drink a lot of alcohol daily, with no apparent mental or physical or social problem at all. Others, the 5%–10%, may also drink a lot, but may end up fighting, being abusive or ill. The same reasoning applies to most drugs or inebriants. 90% of the population may experiment with all kinds of inebriants but then they will moderate their consumption; they will find their limit for consumption, whereby they can continue to work and socialise in a useful, happy way. They do not take too much or too often (which varies enormously between one individual and another), which would result in social dysfunctionality.

The problems arising from the use of drugs[3] in society does not, I suggest, generally derive from the drugs but from the 5%–10% of the population who take them irresponsibly and cause social problems. The focus of regulators should, in my view, be concentrated on helping the 5%–10% with psychological problems, rather than on drugs. Admittedly, as noted above, this is still large number of people. A vital question is perhaps to consider the degree to which we should be concerned to legally prosecute a balanced, caring, successful, productive individual who occasionally becomes intoxicated for pleasure, particularly if periodic intoxication is indeed a biological imperative.

That said, there is also a very useful and important distinction that should be made between psychoactive plants generally and some popular, concentrated chemicals. Siegel (1987) describes several illustrative experiments. In one of them, mice were given access to coca leaf. The mice would nibble the coca leaf periodically. Their behaviour changed but they were still socially functional: eating, grooming, mating, etc. In another experiment, cocaine was made available to the mice. In this experiment, the mice became so addicted that they continued using the cocaine often and repeatedly until they quite quickly died. In another experiment (p. 171) it became readily apparent that the effect on llamas of natural coca leaf and concentrated cocaine were quite different: llamas using cocaine became very aggressive and anti-social, quite unlike the llamas using coca leaf. These brief examples are not provided as proof of the inherent difference between natural and concentrated substances, but merely to add some minimal confirmation of my own observations over several decades.

---

[3] The question of what exactly constitutes a 'drug' is highly complex. See Shulgin (2021) for a comprehensive discussion. As a 'drug' is a substance that changes someone's metabolism or neuro-chemistry, then many substances, such as, for example, radiation, insecticide or even cod liver oil, should also be considered to be drugs. Linguistically, the term 'drug' most probably derives from the Dutch term meaning 'dried goods' in around 1400 CE (Jay 2010:54).

Besides the issue of the personality of the user, it is important to realise that natural, psychoactive plants nearly never cause the extreme social and behavioural problems that are often associated with taking concentrated drugs such as heroin or cocaine. However, I would suggest that, generally—of course, there are exceptions—it is only the vulnerable 5%–10% of the population who would go beyond merely experimenting with concentrated drugs and develop detrimental habits of use. That said, however, although it seems sensible to allow people a degree of freedom to alter their consciousness as they choose, it would not seem prudent to allow crack, for example, to be sold to anyone at corner shops. Careful regulation of such potential transactions is far from simple to resolve. The use of drugs in general is, of course, a highly complex topic, which these brief notes do not adequately or completely explore. I have yet to determine the parameters of what a sensible regulatory system might determine for some commonly used, concentrated chemical drugs, and the degree of availability that might be permitted.

Nevertheless, it is perhaps important to disentangle moral and practical arguments about the potential availability of concentrated drugs and to recognise that the distinctions that most people in contemporary societies make between foods, drugs and medications is somewhat arbitrary and misleading. Two points that could be made in this regard. Firstly, both sugar and television have effects on our nervous systems and brains practically identical to those that could be induced by the consumption of a 'drug'; and it is abundantly apparent that most people in the world have partial or extreme addictions to both sugar and television. It has been estimated that high television use is worse for health than regular tobacco use. High television use is implicated in blood clots and up to a 10% reduction in verbal memory function (Macmillan 2017; Hannah C. 2020). Secondly, to take regularly any form of medication is to take drugs: that they may be legal or desirable for health purposes is irrelevant to the broader moral point.

Concerning specifically aphrodisiac substances and formulas, it could be asked why it is necessary to take them at all. Why can't we just enjoy 'natural' sex without the use of such substances? In answer, firstly, some men, as they age, lose the staying power of youth. Aphrodisiacs can be a great remedy for this issue if used prudently. Secondly, even if full functionality is not impaired in any way, there is, I believe, a natural—one could say almost biological—curiosity in humans to seek the finest or deepest experiences in all domains. We may seek the thrill of dangerous or extreme sports, seek out the best quality music, honey, vegetables, tea, wine or cannabis; we may research the best places in the world to hike or swim; or we may frequent the best libraries, sports events, museums or art galleries that we can find. None of the activities mentioned are necessary; we could just make do with the local, plain and ordinary. But the urge to deepen or enhance experiences, of whatever kind, is, I suggest, a deep and natural feature of the human condition.

3. *Agents of intimacy: psychedelics and aphrodisiacs*

We chase experiences of all kinds; such activity is, I believe—as mentioned above—a feature of the human condition. However, as we all know, experiences, however amazing they might be, come and go. The journey only seems to become deeper, satisfying, relevant and meaningful when we focus on cultivating loving personal relationships and wholeheartedly apply ourselves to useful projects and activities in our daily lives.

In this monograph I focus simply on the aphrodisiac effects of substances. This is somewhat restricted and relatively uninteresting in comparison with the more profound and self-changing potential that psychedelics have to induce mystical experiences, a vast and important topic, which will not be addressed in this publication. That said, deep experiences of sexual ecstasy can come close to—or even be considered to be a form of—mystical experience. However, in my view, the key to all meaningful experiences, whether induced by psychedelics, yogic practices, or other means, is the depth of internal and external intimacy induced.

Nevertheless, a caveat to an over-enthusiastic embrace of aphrodisiacs and agents of intimacy in general, is that having experienced the most exciting sex and the most profound orgasms, there may remain the lingering longing to have such powerful, intimate experiences again. But once a 'peak' has been achieved, subsequent experiences may fall a tad short of the peak experiences of yore. This phenomenon is not, of course, restricted, to the use of psychedelics or aphrodisiacs: the same neuro-logic also applies to achievements in sport, business and performances in general.

Of course, the pleasures of sex are primarily enhanced by the love and emotional intimacy of a partner; however, if one is in a loving relationship, then such pleasures may be significantly enhanced by the judicious use of a range of substances.

I consider psychedelics as agents or tools of intimacy: intimacy with a partner, or with other people, one's environment, the universe at large, and with one's body, mind, sensations and personal biography. Aphrodisiacs may also enhance the sense of intimacy, but only in a more limited way, with one's body and sensations. I believe that it is intimacy in general that is the key that can unlock the blocks and personal strategies of avoidance—sometimes unconscious—that inhibit greater love, understanding and awareness in all areas of human interaction and experience.

Intimacy in general could, in my view, be broadly equated with being 'high'.[4] In a short book, *The Lazy Man's Guide to Enlightenment*, Golas (1978:13–21) argues for a form of pantheism and suggests that one way human interactions can be understood is on a spectrum of energy density. Human consciousness functions in a spectrum of contraction and expansion. At one end of the spectrum is 'mass', where people do not even readily communicate; they are like balls that bounce off each other. At this level of vibration, the contraction of consciousness manifests as pain, fear and unconsciousness. At a higher, more expanded level of vibration we are 'energetic', intensely engaging with each other with passion, anger or energy. At the highest level of vibration, we just like space; being 'unblocked', we are easily able to flow in and out of another's space. This is the dimension of love. Intimacy requires a level of interaction that is like the level of 'space' described.[5] If we are 'high' then we are more intimate with everyone and everything; we may even experience 'becoming' another person, or 'being' the sun or the wind.

---

[4] In the Tantric forms of religion in Asia the connection between female sexual 'energy', as *śakti*, and what might be called 'spiritual' energy is well known.

[5] The tripartite scheme of Golas is also interesting in that it neatly maps part of the metaphysics of the Sāṃkya system of philosophy, which is one of the most influential of all Brahmanical systems of philosophy in India (see Larson 1979). In Sāṃkhya there are two main principles: *puruṣa*, which is pure consciousness; and *prakṛti*, which comprises the mind, senses, the body and the external world. All *prakṛti* is composed of three main qualities (known as *gūṇas*) in varying proportions: *tamas* is dark, black, heavy and inert; *rajas* is red, fiery and energetic; *sattva* is light, white and conscious. This tripartite scheme also applies to foods and people. People are dominated by one or another of the *gūṇas*. I speculate that this scheme may have arisen from observations of natural phenomena: water may be solid (as ice), liquid (as water), or gas (as steam); a fire comprises logs (solid), fire (red and energetic), and smoke.

4. *The brain, yoga, and the moderation of neural chemistry*

Although maybe not everyone would agree, the brain has, in my view, an extraordinary, inherent capacity for self-regulation. By this I mean that we cannot 'cheat' it to gain additional pleasures. If we take a substance to get high that involves no work, then we must pay back afterwards. Stimulants, such as cocaine or ecstasy, or alcohol for that matter, may provide a free ride to a high; but you have to pay proportionally for the high afterwards with a hangover. I don't like any kind of hangover; for this reason, I do not like alcohol, ecstasy or stimulants at all. In contrast, with a decent dose of psychedelics, you usually need to put some work in at the front end; when coming up one may need to process thoughts or feelings, or work through psychological material, which can be difficult. In South America some groups refer to ayahuasca sessions as a *trabajo* (Spanish)/*trabalho* (Portuguese): a 'work'. But having put the work in at the front end, you may then enjoy a long-lasting high afterwards, with not only no hangover, but feeling much better, both mentally or physically, for days or even weeks afterwards.

However, the residual high from psychedelics, which can act as a sacred tonic to the system, may only, I believe, manifest fully if the user has a 'yogic' lifestyle, one that incorporates a routine of disciplined, daily aerobic exercise (particularly outdoors) and the practice of meditation in one of its forms. Breathing strongly and evenly in a context of concentration provides the platform through which, in my view, psychedelic processing is most effective.

The effect of cannabis, which can be mildly psychedelic, is a special case. It can be a highly effective aphrodisiac,[6] especially for women, and may be enjoyed without a significant hangover. But for the hangover to be negligible or even unnoticeable, the user needs to put in 'front-end' work, in the form of daily aerobic exercise, meditation, body-work or yoga. The yogi may enjoy the fullest pleasures and benefits of cannabis.

---

[6] Numerous studies have illustrated the aphrodisiac effect of cannabis. This was observed in Bengal by the Irish doctor and pioneer cannabis researcher William O'Shaughnessy (1843:368), who reported that in many of the patients to whom he experimentally administered cannabis, for a variety of medical complaints, that cannabis was "highly aphrodisiac". Modern sociological and psychological studies, such as those by Tart (1971:141–151) and Earlywine (2002:111–112) confirm the pronounced aphrodisiac effects on most consumers, in terms of heightened arousal, sensation and orgasm. However, there are also reports of cannabis reducing sexual appetite (for example, de Ropp 1957:96). A more recent study (Bhambhvani *et al.* 2020) confirms that cannabis usually increases libido and sexual satisfaction. In general, frequent or even daily use increases the aphrodisiac effect. However, interestingly and consistent with the sometimes ambivalent effects of cannabis, in India some Indian *sādhus* (holy men) who use cannabis in very great amounts—smoking up to 20 or 30 grammes a day—claim that cannabis helps them to maintain celibacy more easily. This could perhaps be explained by some previous research, summarized by Block (2017). Very high-dose THC administered to mice reduced copulatory activity (Dalterio *et al.* 1978; Dalterio 1980); men using marijuana very frequently produce less testosterone (Kolodny *et al.* (1974); and sperm motility appears to be reduced by high-dose THC (Whan *et al.* 2006). However, in another study by Dalterio *et al.* (1977) mice given a low dose of cannabis maintained high levels of testosterone for up to an hour, while testosterone levels dropped sharply for those given very high doses. Bhambhvani *et al.* (2020:2) and Taberner (1985:184) also cite other studies of chronic, heavy use resulting in a decrease of libido. Bhambhvani *et al.* observe that, "These contrasting reports of beneficial and detrimental sexual effects underscore the potential for cannabis to have either a positive or negative impact on sexuality". It seems that if cannabis is used continuously all day, from the time of waking until sleeping, then cannabis may diminish libido. However, even having a partial break in daily use, consuming only in the evening, for example, allows cannabis to continue to have its well-known and extremely enjoyable aphrodisiac effect; and very heavy users who take a break from consuming for a few days or a week may experience exceptional aphrodisiac effects when they resume consumption.

The yogic aspect of cannabis referred to above similarly, in my view, applies to all psychedelics. When Neem Karoli Baba, the guru of Ram Das, was given LSD for the first time in the late 1960s, he famously described it 'yogi medicine'. I fully agree. I believe that the practices of yoga—including breathing exercises, meditation and the discipline of performing postures—are the other side of the coin to psychedelics: I don't think we can fully appreciate the effects and altered states that may arise from psychedelics unless we have a handle on meditation and yoga practices. Likewise, I don't think that we can fully appreciate what yoga is about unless we have had at least one strong dose of psychedelics. A key commonality between yoga and psychedelics is the requirement to put work in at the front end. A fairly common consequence of psychedelic use is the subsequent adoption of some kind of yogic practice.[7]

I do not regard psychedelics as panaceas, though I believe that appropriate use of them can provide profound, beneficial and useful insights into the human condition, psychology and the environment. They are most efficacious when used in conjunction with yogic practice, facilitating deeper insight and discipline. I believe that it is the daily routine of yogic practice which is the vital tool in integrating and processing bodily and mental events and experiences.

5. *Use of psychedelics as aphrodisiacs*

It goes perhaps without saying that the set and setting are absolutely crucial for an enjoyable and insightful psychedelic session. Besides set and setting, another vital aspect psychedelic consumption is dosage. At different dose levels, psychedelics have a wide range of possible effects and it is very important to choose a dose level appropriate to the occasion.

However, it needs to be emphasised that psychedelics are not guaranteed to be aphrodisiac. An unresponsive partner, the prevailing mood and hormonal functioning of the participants, and the setting, all function as contributory factors; and psychedelics may not elicit an aphrodisiac effect at all.

It is also important to note that high or even medium dose psychedelics can be entirely unsuitable and potentially dangerous for someone with a history of mental problems in themselves or in their families, or for people with latent heart problems or a few other medical conditions. Anyone unfamiliar with LSD or other psychedelics should read, before use, a practical guide to consumption, such as by Grof (2019:19–27), Fadiman (2011) or L. S. (2016).

6. *The aphrodisiac effects of stimulant and psychedelic substances*

My experience with psychoactive plants and novel chemicals is relatively limited; but I offer a few personal notes. Although it is not at all the case with everyone I have consulted, my own, very limited experience with MDMA (ecstasy) was that although it increased greatly the desire to have sex, the performance was generally disappointing. I am aware that this is not the case for everyone, though Shulgin (2000:219) comments that most people can't begin to have an orgasm with MDMA. Similarly, stimulants such as amphetamines or *khat* may provide a great sense of intimacy, but in my experience the performance of sex under the influence of these

---

[7] In the 1960s and 1970s, experience with psychedelics—particularly LSD—influenced many people in the USA to take up Buddhist meditation practices (Osto 2019).

substances is also weak and disappointing, or sometimes not possible.[8] Cocaine is excellent for sex, but, as with any stimulant, always comes with a nasty hangover.

Some MDMA users report only a mild hangover or no hangover at all when they first take it; but after a few sessions, hangovers can set in.[9] However, a few experienced users have reported to me that they experience no hangover at all. Hangovers put me off MDMA by 1986, after only two large doses and wonderful experiences. However, as readers may be aware, MDMA is now being used successfully in psychiatric or therapeutic treatment as a useful agent of intimacy—or empathogen—for people who have psychological traumas of various kinds to deal with.

Some of the aphrodisiac plant formulas I tried (see Section 12) were much stronger in effect than others; some produced an almost MDMA-like high. But in accord with my comments above, the greater the stimulant aspect of the high, the greater the hangover.

Myron Stolaroff reports on the therapeutic effects of numerous, new psychoactive compounds, many of which were invented by Alexander Shulgin. Of all the substances considered, Stolaroff (1994:59) records that the effects of 2C-B are highly erotic. This observation is echoed by Shulgin (2000:505), who notes that 2C-B (24 mg) is, "Unbelievably erotic, quiet and exquisite, almost unbearable. I cannot begin to unravel the imagery that imposes itself during the finding of an orgasm". Turner (1994:48) also endorses the "intense and electric" aphrodisiac effect of 2C-B, but he considers mescaline and LSD to be the best psychedelics for an aphrodisiac effect; I agree. I took 2C-B once, in a non-intimate environment. The mild hangover the next day put me off trying the chemical again.

I found *Salvia divinorum* to be aphrodisiac, but I only smoked and ate the leaves a few times as I did not enjoy what I found to be a rather creepy and weird effect, which I did not enjoy very much. However, some people I know very much relish *Salvia.* An experience with peyote buttons in 1979 was remarkable. When making love, my body was 'taken over' and sex became entirely involuntary; my body thrusted on its own, without me 'doing' anything. This was the only occasion in my life when this phenomenon occurred.

It is well known that the effects of LSD, DMT and psilocybin have many effects in common, as they are all tryptamine alkaloids with similar chemistry. They are all capable of inducing classic 'mystical' experiences. As mentioned previously, they can all act as agents of intimacy, facilitating closer contact with oneself, the environment or a partner; they can be powerfully aphrodisiac. In general, all psychedelic/hallucinogenic substances, including also iboga (*Tabernanthe iboga*)[10] and cacti containing mescaline, can be aphrodisiac, in that they facilitate intimacy and focus awareness on the current environment, temporarily by-passing the busy mind.

My own psychedelic preference is and always has been LSD. I have also gained great insights, benefits and divine revelations from both psilocybin mushrooms and ayahuasca. I am of course

---

[8] See Taberner (1985:178–179) who also notes erectile problems with amphetamines. After reviewing the available literature, Abel (1985:63) comments that although there are some reports of increased libido and delayed ejaculation with low-dose amphetamines (for both men and women), at moderate dose levels, maintaining erection and achieving orgasm are impaired.

[9] Interviews with several friends who used MDMA extensively, particularly in the 1990s, anecdotally confirmed this observation.

[10] Seldon (1979:177) notes the native use of iboga in Gabon as an aphrodisiac and as a cure for impotence.

aware that other people have a different point of view and champion 'natural' plant medicines over 'chemicals'. I know several people who are very experienced in psychedelics but who are not big fans of LSD, for various reasons. Terence McKenna, for example, occasionally noted his lack of enthusiasm for LSD, preferring psilocybin mushrooms. However, I merely state my own experiences and preferences after very extensive bioassays on myself.

To date, after over half a century of use, I have experienced no adverse effects at all from any psychedelic or cannabis on my health, mind, memory or well-being; quite the contrary, from LSD particularly I have experienced benefits to health, insight and mental functioning, and obtained important insights into yoga and philosophy in general. I do not believe it is necessary to take such substances frequently to enjoy the benefits. Most people tend to experiment with higher and more frequent doses in youth, use generally declining with age.

Even though psilocybin, DMT and ayahuasca are also potentially and equally aphrodisiac, I prefer LSD because I find the experience of LSD to be smoother than any other psychedelic I have tried. Mescaline is somewhat rare these days, and my experience with it is very limited, so it is not considered here. However, similar observations to those below would hold equally for other tryptamines, were dose levels easier to calibrate.

LSD seems to work best as an aphrodisiac at dose levels between 30 and 50 micrograms. At higher doses, there is the possibility that the trip is too absorbing to allow attention to go towards sex; though it can do. There is also a possibility that either you can't climax, or that you climax almost immediately. Both outcomes can leave you feeling a bit odd. Making love and having an orgasm while high on LSD can be wonderful, allowing temporary access to a kind of heaven, where feelings of love, sound, smell and touch are greatly enhanced. Turner (1994:48) mentions that he likes to have sex near the beginning of an LSD trip. My own preference is to wait until near the end of a trip. I found out long ago that smoking weed or hashish near the end of a trip usually leads directly to excellent sex.

Since I first took LSD, I have regarded it as one of the most useful, important and sacred of human inventions. Even small doses may facilitate the incorporation of the Holy Ghost—or 'possession' (*samāveśa*) by the deity—providing profound insights, a temporary contact with Truth and the Immaculate, and the effect of a general health tonic. In my experience, it is the most effective, predictable, insightful and interesting of all aphrodisiacs, with no side-effects or adverse consequences.

Part 2
Aphrodisiac Plants and Substances

7. *Aphrodisiac plants and combinations*

There are two main aspects to aphrodisiacs: one is an increase in blood flow (particularly to the genitals), the other is the libido effect, the desire to have sex. Some substances increase blood flow, some increase desire, some do both.

During the course of bioassays with various aphrodisiac plants I have noticed that the phenomenon known in cannabis studies as the 'entourage effect' has a parallel in the effects of aphrodisiac plants. Cannabis has dozens of chemicals in various proportions, which vary according to the strain. It was proposed in the late 1960s by the Israeli scientist Raphael Mechoulam, who isolated THC for the first time, that the many cannabinoids, terpenoids and flavonoids in the plant could have an auxiliary psychoactive effect. Taking a particular, individual aphrodisiac plant on its own, even in quite large amounts, may not be as effective as taking a multiple-plant formula. The combination of plants seems to work in most instances much better than any of the individual plants on their own, even when taken in large doses.

A related point is that of dosage. Appropriate dosage is an absolutely essential consideration for all substances, depending on the setting and mood of the consumer. No substance has a 'strong' effect *per se*; it depends on the dose. Likewise, any substance can be dangerous, or not. You can easily die from drinking too much water for a couple of days. Another interesting consideration is that small doses of plants or chemicals, particularly in the psychedelic realm, may be as effective or more useful and enjoyable at low dose levels. Psychedelic substances can trigger intimate experiences, and sometimes low doses may prove more efficacious than larger ones.

This is equally true for individual aphrodisiac plants and combination formulas. Larger doses are not necessarily more effective; small doses may be more effective than larger ones. It is a question of scientifically testing dose ranges of particular plants and formulas in various situations, environments and psychological mood.

8.1 *Individual plants*

There are a vast number of substances that are considered to be aphrodisiac. Many dozens are included in publications by Rätsch and Müller Ebeling (2013) and others. The plants listed below are among the most commonly consumed for aphrodisiac properties. Aphrodisiac recipes from antiquity and the middle-ages in both Europe and Asia sometimes include the most improbable array of ingredients. In traditional Āyurvedic formals (see Section 11.4), we find goat testicles cooked with buffalo meat and cock meat fried in crocodile semen, while Pliny recommends the right lobe of a vulture's lung attached to the body in the skin of a crane, or the right side of an elephant trunk attached to the body with the red earth of Lemnos (Taberner 1985:35), to provide just a few, brief, bizarre examples.

The list below is restricted to the substances that I have either taken or found interesting, most of which are not used in traditional Indian, Āyurvedic formulas. Some of the substances listed below are effective on their own; most of them also feature among other ingredients in various global, aphrodisiac 'combo' formulas.

Corazza *et al.* (2014) note that the use of unlicensed food and herbal supplements to enhance sexual function is greatly increasing. The authors of the study examine the effectiveness and side-effects of yohimbe, maca, horny goat weed, and *Ginko biloba*, which are currently among the most popular aphrodisiacs, used both individually and in combination with other plants.

My comments on effects may be at odds with the findings of other experimenters or consumers. Many factors could account for discrepant reports: my own constitution at the time of consumption, the age of the plant substance consumed (the effectiveness of most plants degrades over time), and the relative quality of the ingredients in the retailed product. So, other people may have had differing experiences (stronger or weaker, for example) from those reported below.

Alstonia scholaris (Dita tree, *saptaparṇī*) is native to both India and China. It has several medicinal applications, is grown for wood and paper, and is known in India as 'the devil's tree' (*śaitān ka jhāḍ*), owing to the toxicity of its leaves, bark and seeds. The crushed seeds of this tree are used as an aphrodisiac. The effect is said to be a tingling in the genitals, prolonging erection and delaying orgasm (Miller 1985:9–11). However, owing to its toxicity, this plant is not included in any aphrodisiac formula known to the author. The pollen of trees planted in Indian cities frequently induces respiratory problems, asthma, sinusitis, fevers and irritation in the eyes among inhabitants (*The Hans India* 2017). I have not tried this plant.

Avena sativa (wild green oats) has been reported to be used as an aphrodisiac (Choueke 1998:72–75). The effect is very mild but the plant may enhance the effect of other aphrodisiacs.

Blue lotus (*Nelumbo nucifera*), used in ancient Egypt, is reputed to have both psychedelic and aphrodisiac effects. Systematic bioassays by myself and a friend—through infusion, smoking and eating the plants—resulted in a pleasant, mildly sedative effect, which was not noticeably either psychedelic or aphrodisiac. This seems to be the general conclusion of those who have tried it (*Erowid* 2021).

Butea superba is a vine-like shrub that grows in South and South-East Asia. It is used as an aphrodisiac, particularly in Thailand; it is effective. Trials did not reveal any noticeable toxicity (Cherdshewasart and Nimsakul 2003).

Cordyceps is a fungus of worldwide distribution of around 600 species, which are parasitic on insects and other arthropods. These days cordyceps is produced commercially in laboratories. Anecdotally, natural cordyceps is far superior to the substance produced in laboratories. The best-known host for one of the species of the parasitic fungus (*Ophiocordyceps sinensis*) is a caterpillar that lives between 3,000 and 5,000 metres in the mountains of parts of Nepal, Bhutan, Tibet and the Kumaun district of Uttaranchal in India. The yellow caterpillar is known as *yarsa/yasha gumba*. Before the rainy season, ghost-moth caterpillar larvae, which live in grassy soil, become infected with the parasite; the caterpillars' bodies then become infused with the fungus. When the fungus reaches the head of the caterpillar it eventually dies a few centimetres underground. From mid-May to mid-June, local people scour the high mountain slopes, scratching away earth in search of the dead caterpillars. The caterpillar tastes like a mushroom and is highly prized by the Chinese, who are the main purchasers in Nepal. *Yarsa gumba* is an excellent stimulant and aphrodisiac and is reputed to be behind the success of

Chinese athletes. This 'Himalayan viagra'—known as *dong cong xia cao* in Chinese, and as *yart swa gun bu* ('herb in the summer and insect in the winter') in Tibetan (Karki and Schaffner 2015)—is added to many products, including teas, alcoholic drinks and anti-aging skin cream. It is expensive: high quality caterpillars retail for $21,120 per kilo in Nepal and for $70,000 in China (Byers 2020); a single, high-quality caterpillar, which is typically a few centimetres in length, costs around $25 in China; and you need around four or five for the full effect.

Catuaba is the name in the Guarani language of South America, which is spoken in parts of Brazil, for the bark of a tree derived from either (*Trichilia catigua* or *Erythroxylum vaccinifolium*). It is a mild stimulant and aphrodisiac and one of the main ingredients in South American multi-plant aphrodisiac formulas, which can be highly effective, available in places such as urban supermarkets in Brazil.

Damiana (*Turnera diffusa*) is a small shrub, primarily native to Texas and Mexico. It also grows in Africa. The leaves can be prepared as a tea. It is not an uncommon ingredient in aphrodisiac formulas. It is a mild aphrodisiac and also has a mild psychoactive effect. Smoking the leaves has a cannabis-like effect (Miller 1985:22).

Dong quai (Chinese ginseng, *Angelica sinensis*) is popular in China as a remedy for female menstrual and menopausal issues. It is also reputed to work well as a female aphrodisiac (Choueke 1998:81–83).

Fo-ti-tieng/Gotu kola (*Hydrocotyle asiatica minor*) is a perennial herb that grows mainly in China and Sri Lanka. It is reputed to lengthen life and is also taken as an aphrodisiac. It can be taken daily as an infusion (Miller 1985:27–32; Seldon 1979:125).

Ginko biloba (maidenhair tree) is a tree native to China. It has been used for 1,000 years in Chinese medicine for a variety of ailments, including dementia, vascular disease, tinnitus and high blood pressure. It is occasionally found in aphrodisiac formulas, though the author has found no noticeable aphrodisiac effect from this plant.

Ginseng, of which there are several varieties, is exported primarily from South Korea. An America species of ginseng (*Panax quinquefolium*) was discovered by a Jewish missionary near Montreal in 1718 (Taberner 1985:68). Ginseng is a well-known stimulant and has aphrodisiac properties, but I have only found this effect to be significant in large doses.

Horny goat weed slightly increases libido and blood flow to the penis. There are many varieties of this plant (*Epimedium*), which is native to China, Japan and Korea. It works as an aphrodisiac, but in my experience the effect is relatively mild.

Maca (*Lepidium myenii*) is also known as Peruvian ginseng. It is native to South America, in the high Andes mountains of Peru. Until the late 1980s it was found exclusively on the Maseta de Bombón plateau close to Lake Junin. The root of the plant, which is highly nutritious, is used for an aphrodisiac effect, affecting libido but not blood flow; the effect is quite mild. There also some reports (unconfirmed scientifically) that maca may reduce the size of the prostate.

Mandrake (*Mandragora officinarum*) is a plant with a long history of use in Egypt, Europe and Asia as both a sedative and aphrodisiac (Manniche 1999:117–119; Schultes *et al.* 2001:90–91). The root or the bark of the root is used. There are several species, which are often confused (Taberner 1985:111–118), one of which, *Mandragora caulescens*, is native to Sikkim and India

(Singh 2006:191). Mandrake is in the *Solanaceae* (nightshade) family of plants, which also includes henbane and belladonna. Henbane and other *Solanaceae* were used as additives to 'magic potions' not only in Asia but also in ancient Greece and medieval Europe (Müller-Ebeling *et al.* 2003:93–96; Hatsis 2014). These plants contain the psychoactive alkaloids atropine, scopolamine (= hyoscine), and hyoscyamine. Mandrake has the highest level of scopolamine amongst these plants. The datura plant also contains scopolamine, particularly in the seeds. Scopolamine is not only psychoactive, inducing hallucinations, but is also aphrodisiac. However, scopolamine is highly toxic and easily causes insanity. It may take many months to recover from a dose of scopolamine; this was my experience from smoking datura seeds. In my view, plants containing scopolamine should be absolutely avoided. I have known people go mad from taking them.

Muira puama (*Ptychopetalum olacoides*) or 'potency wood' is a flowering shrub or small tree indigenous to the Amazon region of South America. A concentrated extract of this plant is commonly used in South American aphrodisiac formulas, in Brazil, Suriname and Guyana. It is only mildly aphrodisiac in my experience.

Tongkat ali (*Eurycoma longifolia*) is a green shrub that grows in South-East Asia, particularly in Indonesia, Malaysia and Vietnam. The root is used. The plant is usually used daily, sometimes in a regimen of two weeks on and one week off (or in other regimens). It boosts testosterone significantly, very noticeably after a few days of use. I have found no testosterone booster more effective than Tongkat ali. While using Togkat ali, aphrodisiac sensations are more or less continuous in the genital region. If other aphrodisiacs are used while 'on' Tongkat ali, then the other aphrodisiacs tend to get boosted. It does not increase blood flow, but engenders excellent orgasms.

Yohimbe, also known as quebrachine, is an indole alkaloid. It is derived either from the bark of an African tree (*Pausinystalia johimbe*) or from an unrelated South American tree (*Aspidosperma quebracho-blanco*). It is available over the counter in the USA and South America but is prohibited in Europe and the UK. Capsules sold in the USA are weak, but those sold in South America are much stronger. Yohimbe has MAOI properties. A single South American capsule has a strong effect and in combination with other drugs can be very dangerous. Some novices report panic attacks, rapid heartbeat, and other negative symptoms. The effects come on very slowly, peaking after about six hours. It gets you high. I used to feel the effects for up to ten days afterwards. Yohimbe is not for the faint hearted. It is highly aphrodisiac; it increases both blood flow and libido. One aspect of yohimbe is that although the aphrodisiac effect is great, orgasm is slightly disappointing, usually barely distinguishable from the preceding, heightened arousal.

8.2. *Individual chemicals*

There are several chemicals frequently reputed to be aphrodisiac, none of which I have tried, with the exception of Shilajeet (which is a natural substance; see below), which I have taken very frequently, both on its own and in combination formulas. In some modern aphrodisiac formulas (see Section 13.1) there seem to be novel chemicals (drilizen and solidilin), which I have not yet been able to properly identify.

Amyl Nitrite is a chemical known as a 'popper' (a name used for a range of chemicals called alkyl nitrites) by recreational users. The vapour from the liquid is usually inhaled from a small bottle. It is a vasodilator (increasing blood flow), thereby lowering blood pressure; it was used as a heart medicine in the 1960s. It can cause dizziness or nausea or skin irritation. Amyl nitrite produces short-lived euphoria and it can have aphrodisiac effects. It was widely used, particularly in homosexual environments, in the 1970s and 1980s. However, in one study, around 30% of male users lost erection. In another study, amyl nitrite reduced penile tumescence (Abel 1985:69).

Bromocriptine (originally marketed as Parlodel) is a dopamine agonist, which blocks the release of prolactin. it has been used to treat Parkinson's disease, type 2 diabetes and other maladies. It is reported by some users, especially women, to have an aphrodisiac effect (Choueke 1998:33, 108–111). It has been used to treat male impotence in haemodialysis patients, but can have unpleasant side-effects, including nausea, vomiting and double-vision (Taberner 1985:160–161).

Flakka is one of the names for the chemical apha-PVP, which seems to be made commercially almost exclusively in China. This chemical belongs to a class of chemicals known as the cathinones, which are similar in structure and function to the active ingredient in *khat*. Alpha-PVP is a strong aphrodisiac but it can also induce hallucinations and paranoia, which may last for days. It is also reputed to be highly addictive. A closely related chemical is alpha-PHP (1-phenyl-2-[1-pyrrolidinyl]-1-hexanone). This was what was the drug that John McAfee was supposedly addicted to for two decades, which sent him quite crazy, until his mysterious death in June, 2021 (Wise 2016).

Ketamine was first isolated (as ketamine hydrochloride) in 1961 by Dr Kalvin Stevens of Wayne State University. It was originally called C1581 and after funding and development by the pharmaceutical company Parke Davis began to be used as an anaesthetic in 1970 (Barillas 2022). Ketamine is drug that is these days used quite commonly recreationally. It is usually either injected or insuflated, but it can be drunk as a liquid or even smoked with tobacco. The disassociation induced by ketamine leads to general inhibition, and at low doses is reported to be aphrodisiac by both men and women. In high doses it can induce unusual experiences of other worlds or realities. Detailed accounts of ketamine experiences were famously penned by John C. Lilly in the last few chapters of his autobiography, *The Scientist* (1997). Sustained use of ketamine leads to renal problems. Overdose can be fatal.

L-arginine is an amino acid that increases the body's production of nitrous oxide. It became popular in the 1980s among athletes and bodybuilders to increased their performance. It also has an aphrodisiac effect (Choueke 1998:102). It is an ingredient in some Indian aphrodisiac formulas.

Phentolamine (sold as Regitine) acts quickly—in about fifteen minutes—as a vasodilator, causing an increase in blood-flow, including the penis. It has been reported to have aphrodisiac effects (Choueke 1998:89–90).

Selegiline/L-deprenyl is used to treat Parkinson's disease. It acts as an MAOI (see Section 5). At lower doses it acts on MAOI-B, increasing dopamine levels in the brain, while at higher doses (more than 20 mg per day) it loses influence on MAOI-B and instead inhibits MAOI-A, which increases serotonin and norepinephrine in the brain. It is reported as an aphrodisiac by some users (Choueke 1998:30, 105–108).

Shilajeet (*śilājīt*) is the dried black/brown extrusion of resin from high-Himalayan rock, widely available in capsule form. It contains numerous minerals and is a very common additive to

Āyurvedic aphrodisiac formulas. I have noticed practically no effects from shilajeet taken on its own. However, I have come to the opinion that shilajeet may have a slight booster or enhancement effect when taken together with other aphrodisiac plants.

Spanish fly (cantharides) is a powder derived from crushing dried, bright green blister beetles (*Cantharis* or *Mylabris genus* or *Lytta vesicatoria*), which are common in the Middle-East, the Mediterranean region, and even Britain. Taken either as a powder or in a tincture, cantharides is a stimulant that has a long history of use as an aphrodisiac, mentioned by Tacitus in his *Annals of Imperial Rome*. It is also used as a blistering agent, but in the British Formulary it is classified as a Schedule 1 poison. A fatal dose is only 32 mg; there are cases of death arising from accidental poisoning with cantharides (Taberner 1985:102–111).

9. *Viagra*

Sildenafil is the chemical in Viagra and Kamagra; tadalafil is a similar chemical used in Cialis; another related chemical, vardenafil, is used in Levitra. Viagra works for about six hours, while Cialis lasts for around forty-eight hours; Kamagra begins working almost instantly. Sildenafil increases blood-flow—notably to the penis— but does nothing to libido.

I tried Viagra once and found it uninteresting as there was no erotic component to the experience. Alarmingly, six hours after taking it I smoked a joint. The effect was extremely unpleasant—a mild bad trip—and I had to lie still and concentrate for quite a long time; another joint two hours after that had the same effect. Sildenafil is, in my experience, the only drug I have ever taken that is incompatible with cannabis. More seriously, sildenafil does weird things to my heart. Three days after taking sildenafil I always got mild cardiac arrhythmia and other alarming sensations in my heart. I found out about this not directly from my single Viagra experience, but because sildenafil is sometimes secretly added to aphrodisiac formulas (and not listed as an ingredient); and every time I took any such formula that included sildenafil I had the same experience. A clue to the addition of sildenafil is that you see blue light. I quite quickly came to the conclusion that sildenafil should only be taken infrequently and with great caution.

One of the most effective aphrodisiac formula I ever tried was Libidus (Section 13.2, no. 43). Both libido and erection were exceptional. However, irregular heart activity three days afterwards made me suspect the addition of sildenafil, which is not listed as an ingredient. I subsequently found out that sildenafil is indeed secretly added to the Libidus formula (Canada Vigilance Program 2013).

Part 3
Indian Aphrodisiac Formulas, Ancient and Modern

10.1 *A note on spelling and pronunciation of names of plants native to South Asia*

In this and the following sections numerous plants native to South Asia are mentioned. Most of them have many names, in not only local languages, but also in Hindi and Sanskrit. I have attempted to standardise the spellings of names; however, minor discrepancies persist. The names are rendered primarily in either Hindi or Sanskrit, as that is the most common way names of plants are displayed on the labels of both single and multi-plant formulas available in India.

Without specifying the pronunciation of all Hindi or Sanskrit phonemes, the following may be noted:

ā = 'long' a (as in 'car')
ī = 'long' i (as in 'easy')
ū = 'long' u (as in 'ouse')
c = ch (Caraka = Charaka)
ś = sh
ś = sh (but slightly compressed)
ṃ = nasalised n

10.2 *Aphrodisiac formulas used by the Mughals in India (1526–1803)*

The Mughals enjoyed using aphrodisiacs. Nath (2008:142–144) lists several formulas and remarks that a *hakīm* (a doctor trained in Unani medicine) in Agra still prepares these them. Usually prepared as an ointment or paste, one group of these plant concoctions was applied to the penis and inside the vagina, which said to provide great pleasure to women, with little effort required by the man. Another class of concoctions (*stambhana*) is used to delay ejaculation (Nath 2008:140–142).

In another group of concoctions, to boost libido and pleasurable experience, are the aphrodisiac formulas, which are listed by Nath, as follows. All the plant formulas are taken with milk, honey (or sugar) and usually *ghī* (clarified butter).

1. a concoction of *utaṅkan/uccaṭā/uccāṭa* (crab's eye/rosary pea/jequirity bean, *Abrus precatorius*) and *muleṭhi* root (liquorice, *Azadirachta indica*).
2. powdered *muleṭhi.*
3. the testes of a he-goat or ram boiled with milk and sugar, also mixed with *pīpal* (*Ficus religiosa*) and salt or black sesame seed.
4. the powder of *vidārīkāṇḍ/vidārī/kudzu* (*Hedysarum tuberosum/Pueraria tuberosa*); *vāṃslocan/bāṃslocan* (a white powder made from nodal joints of some species of bamboo, containing mainly silica and water, with traces of lime and potash); and the seeds of *ātmagupta/kauṃc/kevāñc/kapikacchu* (cowhage, *Mucuna pruriens*).
5. the powdered seeds of either *kauṃc* or *bāṃslocan,* and *tāl makhānā/kokilakṣa* (fox nut/prickly waterlily, *Asteracantha longifolia*).

6. the powder of *ciraumjī* (*cārolī*, almondette kernels, *Buchanania latifolia/lanzan/cochinchinensis*); *gulmohar* (*muhar/morta/murahari*?), flame tree/peacock tree/fragrant royal poinciana, *Delonix regia*; and white *vidārīkānd*, taken with milk for fifteen days.
7. a concoction of *singhāḍā* (water caltrop/Indian water chestnut, *Trapa bispinosa*); and *kaserū* (bullrush, *Scirpus grossus*), with fruits of *mahuā* tree (*Madhuka longifolia*), and *kṣīrakākolī* (*Lilium polyphylum*), prepared on a light heat.
8. rice pudding (*khīr*) prepared with the de-skinned *urad* pulse.
9. *khīr* prepared with the eggs of the *caṭak* (sparrow).
10. a concoction of *mulethi* and *mulahri* (= gulmohar?).
11. *śatāvarī/satvīrya* (*Asparagus racemosus*), *gokhrū/gokṣurā* (puncture vine/caltrop, *Tribulus terrestris*) and *pīpal* (*Ficus religiosa*).
12. powdered *nāgaurī/aśvagandhā* (*Withania somnifera*) and *vidārīkānḍ* (*Hedysarum tuberosum/Pueraria tuberosa*).
13. powdered *āmalakī* (*āṃvlā, Emblica officinalis/Phyllanthus emblica*).

11. *Aphrodisiac plants used in Āyurveda: complex formulas, ancient and modern*

In this section we will survey the use of several plants that are most widely used in Āyurveda as aphrodisiacs, comparing their use in traditional Āyurvedic treatments and in modern formulations.

The South Asian medical tradition of Āyurveda ('knowledge of life-span') acknowledges three primary, historical authorities, namely Caraka, Suśruta and Vāgbhaṭa, the 'great three' (*bṛhat trayī*).[11] The earliest formulations of the compendia (*saṃhitās*) of both Caraka and Suśruta date to the early centuries BCE; these medical texts attained, after additions, their current form in the early centuries CE. Vāgbhaṭa's *Aṣṭāṅga Hṛdaya* ('heart of medicine') largely synthesises the treatments detailed by Caraka and Suśruta and was probably composed around 600 CE (Wujastyk (1998:40, 105, 238). It became one of the most widely referred to medical texts in Asia. Later Āyurvedic authorities built generally on the treatments contained in these texts, which included oleation, fomentation, emesis, purgation and other treatments.

In traditional Āyurveda it is believed that having a good sex-life is important for health (see below). The three foundational treatises on Āyurveda, by Suśruta and Caraka and Vāgbhaṭa all have sections on aphrodisiacs, known as the science of *vājīkaraṇa*, which is one of the eight branches of Āyurveda. This science still flourishes in India and several modern formulas are listed in Section 12.

11.1. *The eight branches of Āyurveda*

Āyurvedic treatments (*cikitsā*) are traditionally organized as eight branches (or lotus petals) of medicine, namely:

kāyacikitsā (internal medicine for the body)
śalyatantra (surgery)
śālākyatantra (ear, nose, throat and cephalic diseases)
bālā/kaumārabhṛtya (children's ailments/paediatrics, obstetrics, gynaecology)
agadatantra/viṣagaravairodhatantra (toxicology)

---

[11] This tripartite classification is relatively recent, perhaps dating from the 18th century.

graha/bhūtavidyā (psychiatry/demonology)
rasāyana (rejuvenation)
vājīkaraṇa (aphrodisiacs)

11.2. *The importance of sex in Āyurveda*

In Āyurvedic understanding it is generally maintained that a satisfying sex life is important for health; hence an entire branch of treatment (*vājīkaraṇa* = 'stallion capabilities/actions') is concerned with this topic and the use of aphrodisiac plants, animals and minerals to enhance sex experience and semen production. This recommendation is primarily based on the general notion in Āyurveda that no natural urge should be supressed, including those related to urine, faeces, semen, flatus, vomiting, sneezing, eructation, yawning, hunger, thirst, tears, sleep and breathing after exertion. Suppression of these urges is said to lead to health disorders. However, even if accustomed to it, a wise person avoids excessive sexual intercourse (Caraka I.VII.3–4, 34).

In the section on aphrodisiacs in his compendium, Caraka maintains (Caraka II.II.3–15) that a conscious person should use aphrodisiacs regularly because virtue, wealth, pleasure and fame depend on it; and also the production of male progeny, which is an issue of great concern in traditional India. Caraka also notes that the foremost aphrodisiac for a man is an exhilarating woman; and that the qualities in a woman, such as a good appearance, improve when she finds a suitable man.

11.3. *Aphrodisiac formulas in South Asia*

Sood *et al.* (2005) detail around 800 plants that can have an aphrodisiac effect; and that publication only details plants native to India! Multi-plant aphrodisiac formulas are also produced in Africa, South America and other parts of Asia, particularly in China. The author has explored some of these formulas, which mostly use plants that are different to those used in India, and which are often not native to South Asia; however, as mentioned previously, non-South-Asian formulas are beyond the scope of this short book. Numerous aphrodisiac formulas are also provided in dozens of Āyurvedic and alchemical texts that were composed later than those of Caraka and Suśruta, but the survey in this book, apart from a few notes, is restricted to the formulas found in the foundational texts of Āyurveda and in modern use.

11.4. *Aphrodisiac formulas in Caraka*

Caraka (II.II) describes, in a chapter of four sections, around thirty-five aphrodisiac formulas, which contain numerous plants and animal products. These concoctions are variously heated and cooked in ghee and other substances. It may perhaps be surprising to learn how many plants can have an aphrodisiac effect, to a greater or lesser degree.

The aphrodisiac formulas provided by Caraka in his compendium (for a useful, brief summary, see Dalal *et al.* (2013), which are said to provide the strength of a horse or a bull, contain numerous ingredients, some as bizarre as it gets. Some of the plants listed, such as sugarcane, are not aphrodisiac in themselves; others are well known as aphrodisiacs, and a few are of uncertain botanical identity. Caraka also observes (II.II.2.20–25) that, "One gets stimulated like a bull by massage, anointing, bath, perfumes, garlands, adorations, a comfortable house, bed and chairs, untorn favourite clothes, the chirpings of favourite birds, the tinkling of ornaments of women, and the gentle pressing of the body by favourite women and others".

The first formula (II.II.1.24–32) is as follows (all plant ingredients are listed in English in Section 8 below).

> *śara* (roots), *ikṣu* (roots), *kāṇḍekṣu, ikṣuvālikā, śatāvarī, payasyā, vidārī, kaṇṭakāri[kā], jīvantī, jīvaka, medā, virā, ṛṣabhakā, balā, ṛddhi, gokṣuraka, rāsnā, kappikacchū, punaranavā.* 120 gms of each of these drugs, mixed with 2.56 kg of new, black gram should be cooked in 10.24 litres of water till a quarter of it remains.
>
> Then a paste of *madhuka, drākṣā, phalgu, pippalī, kappikacchū, madhūka,* and *śatāvarī* should be added to it, along with the juice of *vidārī, āmalaka* and *ikṣu*, added separately, along with 2.56 kg of ghee and 10.24 litres of milk. This should be cooked until only ghee remains. This should be filtered well and added to with 640 gms powdered sugar, 640 gms of *vaṃśalocana*, 160 gms of *pippalī*, 40 gms of *marica*, 20 gms of *tvak*, 20 gms of *elā*, and 20 gms of *nāgakeśara*. 320 gms of honey should be added. Then boluses of 40 gms each should be made from the concoction.

Caraka's second formula (vv. 33–37) is for an aphrodisiac ghee.

> 2.56 kg of both newly harvested grains of black gram and seeds of *kappicacchū* should be boiled together with 160 gms each of *jīvaka, ṛṣabhaka, virā, medā, ṛddhi, śatāvarī, madhuka* and *aśvagandhā*. Then, 640 gms of ghee, 6.4 litres of cow-milk, and 640 mls of both the juice of *vidārī* and *ikṣu* should be added and cooked. To this ghee preparation, 160 gms of sugar, *vaṃślocana*, honey and *pippalī* should be added.

The formulas that follow are for various kinds of meat soup (*piṇḍarasa*). In the first formula (vv. 38–41), the meat used is either that of a cock, peacock, partridge or goose. The only aphrodisiac plant used in this formula is *vaṃslocana.* The next concoctions comprise: goat's testicles cooked with buffalo meat; sparrow meat; cock meat fried in crocodile semen, and the eggs of fish, goose, peacock and cock.

In the second section of Caraka's chapter on aphrodisiacs, more formulas are described, which contain similar ingredients to those described above. Besides *ṣaṣṭika* rice, honey, ghee and cow-milk, the first formula provided contains *kapikacchū, balā, mugdgaparṇī, jīvantī, jīvaka, ṛddhi, ṛṣabhaka, kākolī, gokṣura, madhuka, śatāvarī, vidārī, drakṣā, kharjūra* and *vaṃṣalocana.*

In order to make the second formula, one should collect the semen of sparrows, geese, cocks, peacocks, tortoises and crocodiles (the mind boggles), which should be consumed together with the fat of the *kuliṅga* bird (a kind of sparrow).

The formulas that follow are similar in form, comprising nearly all the same plants noted above. In one of them, the milk of a cow fed on black gram leaves, or sugarcane, or *arjuna* leaves is said to be aphrodisiac. Another recommends *śapharī* (fish) *rohita* (fish) and goat's meat soup. One formula comprises only *śatāvarī*, milk, honey, sugar and *pippalī* (long pepper).

The plants used in the other formulas in this chapter that have not already been mentioned are *śṛṅgāṭaka, mṛdvikā,* and *māṣa.*

11.5. *Suśruta's aphrodisiac formulas*

Suśruta, in his chapter on aphrodisiacs (Suśruta, vol. 2, XXVI), first details the causes of impotence—such as old age, a congenital condition, forced intercourse, excessive sexual activity, and voluntary continence—and then provides formulas similar in style and content to those of Caraka.

Among the recommendations (vv. 9–13) are the testes of a goat or of a porpoise, eaten together with salt and *pippalī*; the testes of alligators, mice, frogs and sparrows; or the eggs of a tortoise, alligator or crab, or the semen of a buffalo, ass or goat.

Some of the plants that Suśruta includes in his formulas are also in Caraka's formulas, while some are not. Like Caraka, Suśruta includes *vidārī, āmalaka, gokṣura* and *ātmaguptā* (= *kappicaccū*) in his formulas. However, Suśruta also includes the sprouts, bark, roots and fruit of the *aśvattha* (peepal) tree, the seeds of *kokilakṣa*, and powdered *uchchatā.*

11.6. *Complex plant formulas in traditional medicine*

As explored by Clark (2021:21–22), in the traditional medical systems of many cultures of the world, remedial plants are very often taken as multi-plant formulas. In Asia multiple plant formulas have for at least 3,500 years been used to engender psychedelic experiences (Clark 2020), similar in effect to ayahuasca, the plant formula used in South America and, since the 1980s, increasingly in the wider world. The use of multi-plant formulas for medicinal, therapeutic, visionary or psychedelic experience is different to the approach in modern allopathic medicine, in which plants known to be used or effective in traditional cultures as remedies are usually analysed and tested with the aim of isolating a particular molecule or chemical ingredient that is most effective in the treatment of a malady.

The isolated ingredient, which is generally concentrated and more potent than in its plant form, is then marketed as a drug, which may be manufactured in combination with other isolated ingredients from other plants or with a synthetic chemical. One of the factors in the manufacture and sale of patented drugs is, of course, the financial gain that accrues; far more than would be derived from the sale of a natural plant remedy.

In contrast, in most traditional cultures effective medicinal plants are very often used in combination with other plants. In many traditional remedial formulas not only is the full spectrum of the chemical ingredients of the medicinal plant consumed but additionally the full spectrum of ingredients of other plants. Plant combinations may have a synergetic effect more potent than that derived from a single plant.

For example, when the Prussian naturalist Robert Hermann Schomburgk was trying in 1937 to identify the plant used in South America for making the poison *curare*, applied to the darts of poison arrows, it was originally believed that the essential poison of *curare* derived from just one liana, *Strychnos toxifera*, and that the admixture plants used by natives to make the poison were redundant. This view subsequently proved to be incorrect: the additive plants have a synergetic effect, increasing the potency of the *curare* (Plotkin (1993:155–156). The great botanist and explorer Richard Evans Schultes would eventually identify and document more than seventy species of plants employed to make *curare* in the Amazon region (Plotkin *et al.* (2017:102).

In Āyurvedic medicine in South Asia it is apparent that for the treatment of most maladies, complex plant formulas are recommended. If we peruse the many dozens of medical formulas provided by Caraka, Suśruta and Vagbhāṭa, it is readily apparent that for a host of medical issues remedies may comprise four or five or even up to fifty or more plants or their extracts.

These remedies are used in formulas for oleation, fomentation, emesis, purgation and other treatments.

The same principle applies to Āyurvedic aphrodisiac formulas. As indicated above, nearly all of the aphrodisiac formulas given by the classical Āyurvedic authorities are multi-plant formulas. Individual plants taken even in large amounts may be much less effective than multiple-plant formulas that may utilize up to around fifty different kinds of plants in small amounts.

11.7. *The aphrodisiac plants used in traditional Āyurveda*

The animal ingredients of the aphrodisiac formulas discussed by Caraka and Suśruta have never as far as I am aware ever been analysed properly in the modern era for aphrodisiac potential. However, most of the plants used in modern Āyurvedic aphrodisiac formulas use those used in the formulas detailed by Caraka and Suśruta. Some of the plants that comprise early aphrodisiac formulas are psychoactive and are also used in South Asia in multi-plant psychedelic formulas (Clark (2021).

All of the plants recommended by Caraka and Suśruta are listed below.[12] Even though featuring in their formulas, some of them are common foods or spices that currently have no psychoactive profile or known aphrodisiac effect. It could be that some of the plants may have properties yet to be discovered, or perhaps they may act synergistically with the other ingredients. Several of the plants have multiple synonyms and are of uncertain botanical classification; so there may possibly be mistakes in the botanical classifications identified below.

Plants with a known (perhaps slight) or greater aphrodisiac effect are marked with an asterisk.

āmalaka: *āṃvlā, Emblica officinalis/Phyllanthus emblica*
**aśvagandhā*: *Withania somnifera*
**aśvattha*: peepal tree
**balā*: *bījband/khareṭī*, country mallow, *Sida cordifolia*
drākṣā: grapes, *Vitis vinifera*
elā: cardamom, *Elettaria cardamomom*
**gokṣuraka*: *gokhrū/gokṣurā/gokṣura*, *Tribulus terrestris*
ikṣu (roots): sugar cane
ikṣuvālika: wild sugar cane/Kans grass, *Saccharum spontaneum*
**jīvaka*: orchid, *Crepidium acuminatum/Malaxis acuminata*

---

[12] Many subsequent Āyurvedic treatises on aphrodisiacs largely replicate the formulas and ingredients specified by Caraka and Suśruta. For example, in the *Rāja Mārtaṇḍa* (see Sastry 2010) of King Bhoja (1010–1055 CE) there is a section on elixirs and aphrodisiacs (ch. 33, pp. 134–139), in which twelve multi-ingredient elixirs are detailed. The first two formulas are for good fortune, while the third is for the promotion of intelligence; the other nine formulas are described as elixirs. The ingredients of these nine formulas comprise many identical ingredients; honey and *ghī* are in most of them. Other ingredients are: seeds of the *palāśa* tree (*Butea monosperma*); *viḍaṅga* (*Embelia ribes*/white-flowered embelia); the fruits of *āmalaka*; oil from the seeds of both *aṅkola* (*Alangium salvifolium*) and *sitasarṣarpa* (white mustard), applied in the nostrils; root of *kuṣṭha* (*Saussurea lappa*); *aśvagandhā* (*Withania somnifera*); *bhasm*s (see Section 11.9) of iron, mica, mercury, *śilājīt*, lodestone/magentite (*kāntaloha*); *cakrāṅga* (*Cocculus tomentosus/Abuta rufescens*) [I believe that Sastry's identification of *cakrāṅga* with either *Cyperus rotundus* or *Tinospora cordifolia* may be incorrect]; *viṣa* ('poison'), identified by Sastry (p. 137) as *Aconite heterophyllum/ atīs* root (though not attested in dictionaries); *vānarī* (seeds of *Mucuna pruriens*/cowhage?); *īkṣura* (= *kokilakṣa/Asteracantha longifolia*); seeds of *mayūra* (*Achyranthes aspera* [in the amaranth family]); *nistuṣikrutayava* (a kind of barley); *māṣa* (mung bean/*Phaseolus munga*); *śāli* (red rice); *godhūma* (wheat); *māgadhikā* (= *pippalī*/long pepper).

**jīvantī*: orchid, *Desmotrichum fimbriatum*
**kākolī*: *Roscoea purpurea*
kāṇḍekṣu: wild sugar cane/Kans grass, *Saccharum spontaneum*
**kaṇṭakāri[kā]*: yellow-fruit nightshade,
Solanum virginianum/surattense/xanthocarpum
**kappikacchū*: *ātmagupta/kauṃc/kevāṃc*, cowhage, *Mucuna pruriens*
**kokilakṣa*: *tāl makhānā*, fox nut/prickly waterlily, *Asteracantha longifolia*
kharjūra: dates
kṣīṛikā: milk, rice and ghee
**madhuka*: *mādhūkā/madhukā/madhūka*, Mahuā tree, *Madhuka longifolia*
marica: black pepper, *Piper nigrum*
māṣa: *urad*, black gram, *Phaseolus mungo*
**māṣaparṇī*: *Teramnus labialis*
medā: *mahāmedā, Polygonatum cirrhifolium/verticillatum*
mṛdvikā: grapes, *Vitis vinifera*
mugdgaparṇī: *Phaseolus trilobos*
**nāgakeśara*: Assam ironwood tree, *Mesua ferrea*
payasyā: *Holostemma rheedianum*
**phalgu*: redwood fig tree, *Ficus hispida*
pippalī: long pepper, *Piper longum*
**punaranavā*: *Boerhavia diffusa*
**rāsnā*: *Pluchea lanceolata*
**ṛddhi*: *Lantanthera edgeworthii* or *Habenaria intermedia* [this plant has sixty-six synonyms in Sanskrit]
**ṛṣabhaka*: *Dienia/Malaxis/Microstylis muscifera*
śāli: a kind or rice
śara (roots): a reed/Baruwa grass, *Saccharum sara/bengalense*
**śatāvarī*: *Asparagus racemosus*
śrāvaṇī: *Sphaeranthus indicus*
**śṛṅgāṭaka*: *Trapa natans/bispinosa*
tvak: cinnamon, *Cinnamomum zeylanicum*
**uchchatā*: *nāgaramustā, Cyperus scariosus*
vaṃśalocana: pith of bamboo, *tugākṣiri, Bambusa arundinacea*
**vidārī*: *kudzu, Hedysarum tuberosum/Pueraria tuberosa*
**virā*: = [probably] *kākolī, Roscoea purpurea* [this plant has fifteen synonyms in Sanskrit]

11.8. *Modern aphrodisiac formulas*

If you enter any pharmacy in India you will see in most of them, displayed on shelves and in cupboards, numerous packets and bottles of so-called 'Āyurvedic' aphrodisiac formulas, some comprising four or five ingredients, other containing up to around fifty ingredients. These formulas mostly comprise plant extracts but some include various metals, minerals and shells, which began to feature more significantly in Āyurvedic treatments about 1,000 years ago, as an overlap began between the two Āyurvedic branches of *vājīkaraṇa* and *rasāyana* (Wujastyk (2017:27), which is the science of longevity. In *rasāyana*, shells, minerals and metals are commonly used. There are many dozens of these formulas on the market, which come and go over the years, due to their popularity (or otherwise), and depending on the vagaries of the business world.

Over the course of twelve years, between 2005 and 2017, I purchased around forty different, Āyurvedic, aphrodisiac formulas, in the form of either capsules, powders or pills (see Section 11). Some formulas contained substances such as opium, which are technically illegal. There are formulas for both men and women, though more formulas for men are available than for women. In classical Āyurveda, the formulas provided are generally for men, even though many plants would affect women also. I recorded the ingredients of thirty-two products made and distributed throughout India. Bioassays and anecdotal reports revealed that some were not very effective, others were found to be too strong (inducing excessive palpitations and sometimes a significant hangover), and some were highly effective, leaving little or no hangover. Overall, after scientific evaluation, it was found that common ingredients of most modern Indian Āyurvedic aphrodisiac formulas are not toxic or particularly dangerous to health (Rathva *et al.* 2017).[13]

Many factors influence the perceived potency of the products, including not only the usual range of effects of particular plants, but also the constitution of the consumer at the time, the 'setting' in which the product is consumed ('romantic' or otherwise), the age of the organic ingredients, which degrade in potency over time, and the specific dosage included in a particular formula.

Perusing the ingredients of the thirty-two formulas under consideration, it is evident that all of the plants recommended by Caraka's and Suśruta are included in modern aphrodisiac formulas. The most frequently mentioned, effective, aphrodisiac plants that are recommended by Caraka and Suśruta, and which are also included in modern formulas are as follows.

aśvagandhā: *Withania somnifera*
balā: *bījband/khareṭī*, country mallow, *Sida cordifolia*
gokṣuraka: *gokhrū/gokṣurā/gokṣura*, *Tribulus terrestris*
jīvaka: orchid, *Crepidium acuminatum/Malaxis acuminata*
jīvantī: orchid, *Desmotrichum fimbriatum*
kākolī: *Roscoea purpurea*
kappikacchū: *ātmagupta/kauṃc/kevāṃc*, cowhage, *Mucuna pruriens*
kokilakṣa: *tāl makhānā*, fox nut/prickly waterlily, *Asteracantha longifolia*
madhuka: *mādhūkā/madhukā/madhūka*, Mahuā tree, *Madhuka longifolia*
māṣaparṇī: *Teramnus labialis*
medā: *mahāmedā*, *Polygonatum cirrhifolium/verticillatum*
nāgakeśara: Assam ironwood tree or cobra saffron, *Mesua ferrea*
śatāvarī: *Asparagus racemosus*
śrāvaṇī: *Sphaeranthus indicus*
śṛṅgāṭaka: *Trapa natans/bispinosa*
uchchatā: *nāgaramustā*, *Cyperus scariosus*
vidārī: *kudzu*, *Hedysarum tuberosum/Pueraria tuberosa*
virā: [probably] *kākolī*, *Roscoea purpurea* [this plant has fifteen synonyms in Sanskrit]

Many modern formulas also include aphrodisiac plants not mentioned by Caraka and Suśruta. Some of the many dozens of plants included in modern aphrodisiac only appear once; others are more frequently included. The most common of these are:

---

[13] Rathva *et al.* (2017) tested: *Withania somnifera*, *Clorophytum tuberosum*, *Mucuna puriens*, *Blepheris edulis*, *Puraria tuberosa*, *Myristica fragrans*, *Hygrophilia spinosa*, *Piper cubeba*, *Cinnamomum camphor*, *Syzygium aromaticum*, *Zingiber officinalis*, *Piper nigrum*, *Piper longum*, and *Crocus sativus*.

akarkarā/ākārakarabha (Spanish chamomile, *Anacyclus pyrethrum*)
babūl goṇd (thorny acacia tree, *Acacia arabica/nilotica/Vachellia nilotica*)
giloy, guḍūcī (*Tinospora cordifolia*)
harītaki (black myrobalan tree, *Terminalia chebula*)
jāyphal/jatiphal/jāvitrī (nutmeg, *Myristica fragrans*)
kuclā/kucilā/kupīlu (strychnine tree, *Strychnos nuxvomica*)
lāl mūslī/semal/śālmali/kapok (silk cotton tree, *Salmalia/Bombax malibaricum/ceiba*)
safed (white) *mūslī* (*Chlorophytum arundinaceum/borivilianum*)
sālam/sālab miśri (*Orchis mascula/latifolia*)
samūdraśoś/samūdra patra/vidhārāv/vṛddhadārū (Hawaiian baby woodrose/elephant creeper, *Argyreia speciosa*)

Somewhat curiously, *āṃvlā* (*āmalaka*, emblic myrobalan/India gooseberry) is a plant frequently referred to by Caraka in his *rasāyana* formulas, followed by the other myroblans: the chebulic and belleric myrobalans (Wujastyk 2015:57); it is also included in both ancient and modern aphrodisiac formulas. Although tests have shown the plant to have aphrodisiac properties (Pathak *et al.* 2011), consuming it does not in my experience lead to any noticeable effect in that regard.

11.9 *Metals and minerals*

Also very commonly included in modern aphrodisiac formulas are various minerals and metals, in particular shilajeet/*śilājīt* (*Asphaltum punjabianum*), a kind of tar that exudes from high altitude rocks (see 9.2). Other additives include various shells, metals and *bhasm*s (powders):

gold, mercury, sulphur in a ratio of 1:8:24 (*makardhvaj*)
calcinated iron ash (*lauh bhasm*)
calcinated mica ash (*abhrak bhasm*)
calcinated silver ash (*raupya bhasm*)
calcinated gold powder (*svarṇ bhasm*)
calcinated tin powder (*vaṅg/baṅg bhasm*)
calcinated copper and iron sulphide (*suvarṇāmakśak bhasm*)
calcinated pearl infused in rosewater (*mukta piṣṭī*)
tin, lead, zinc: processed in aloe vera juice and turmeric (*trivaṅg bhasm*)
gold foil (*svarṇ varaq*)
mercury (*pārad/pārā*)
yellow sulphur/brimstone (*gandhak*)

Makardhvaj is used as a traditional medicine in the treatment of sexual dysfunction in the rural population of South Asia (Roy *et al.* (2017) and is the most commonly used *bhasm* in modern Āyurvedic formulas. The medicinal effectiveness of minute particles of gold, one of the main ingredients of *makardhvaj*, has been explored (Panyala *et al.* (2009). Outside the South Asian context, several metals and minerals (phosphorus,[14] zinc, calcium, silicon, sodium, potassium, sulphur, magnesium, manganese, selenium and vanadium) are included as aphrodisiac potentiators in some publications (for example, Lee and Lee 1994:199–201). In South Asia, besides *makardhvaj*, the most commonly prescribed aphrodisiac metal is zinc, which has been found to increase sperm production (Fallah *et al.* 2018).

---

[14] Phosphorous was included in some aphrodisiac formulas produced in the late 18th and 19th centuries; but it is highly toxic and caused cases of poisoning (Taberner 1985:83–85).

11.10 *The 'top fourteen' aphrodisiac plants used in modern formulas*

From the dozens of ingredients used in modern, Āyurvedic, aphrodisiac formulas, listed below are the 'top fourteen' plants that most are most commonly included.

1. *akarkarā/ākārakarabha*, Spanish chamomile: *Anacyclus pyrethrum*
2. *aśvagandhā*: *Withania somnifera*
3. *balā/bījband/khareṭī*, country mallow: *Sida cordifolia*
4. *gokṣura*: *Tribulus terrestris*, puncture vine, caltrop
5. *giloy/guḍūcī*: *Tinospora cordifolia*
6. *jāyphal/jatiphal/jāvitrī*, nutmeg: *Myristica fragrans*
7. *kappikacchū/ātmagupta/kauṃc/kevāṃc*, cowhage: *Mucuna pruriens*
8. *kesar*: saffron, *Crocus sativus*
9. *kokilakṣa*: *tāl makhānā*, fox nut/prickly waterlily: *Asteracantha longifolia*
10. *safed* (white) *mūslī*: *Chlorophytum arundinaceum/borivilianum*
11. *sālam/sālab miśri*: *Orchis mascula/latifolia*
12. *samūdraśoś/samūdra patra/vidhārāv/vṛddhadārū*, Hawaiian baby woodrose/elephant creeper: *Argyreia speciosa*
13. *śatāvarī*: *Asparagus racemosus*
14. *vidārī*: *kudzu*, *Hedysarum tuberosum/Pueraria tuberosa*

Most of the plants recommended by Caraka as aphrodisiacs and those listed above have been found to have aphrodisiac properties after scientific tests were conducted.

gokṣura: see Kotta *et al.* (2013).[15]
kesar: see Kotta *et al.* (2013).
kokilakṣa: see Chauhan *et al.* (2011).
safed mūslī: see Kotta *et al.* (2013).
śatāvarī: has been found to significantly increase orgasm (Chate *et al.* 2017).

11.11. *Conclusion*

From the preceding discussion, it can be seen that there is considerable overlap between the plants recommended by Caraka and Suśruta for aphrodisiac effect, and those most popular in modern formulas. However, it is also evident that most modern formulas use some plants that are not mentioned by the classical Āyurvedic authorities, possibly because they were not aware of those plants, or possibly because those plants were introduced to South Asia after the time of Caraka and Suśruta, around 2,000 years ago.

Some of the plants used in Āyurveda as aphrodisiacs are not recorded in modern aphrodisiac compendia (for example, Lee and Lee 1994; Rätsch and Müller Ebeling 2013); more research will undoubtedly reveal currently unknown aphrodisiac properties of several plants native to South Asia. As mentioned previously, a combination of plants can be far more effective than the use of an individual plant, even when taken in large amounts. There is still a great deal to learn about the effective chemistry of numerous substances on the human organism.

[15] Stuart (2021) notes that this plant may have adverse effects on the male prostate gland.

12. *Contemporary Āyurvedic aphrodisiac formulas (powders, capsules, pills)*

Below is listed a selection of commonly available, inexpensive, Indian, Āyurvedic, aphrodisiac preparations, all of which I have experimented with. The listed ingredients and quantities are presented below as stated on the labels of the packets. Spellings and botanical identities of the ingredients have been broadly, but not perfectly, harmonised in either Hindi or Sanskrit, as most of the formulas, as previously noted, are specified in these languages. Some formulas are more effective than others.

12.1 FOR MEN

1. Baidyanath Vita-Ex Gold (Shree Baidyanath, Jhansi)

1. *ātmagupta/kauṃc/kevāṃc/kapicacchu* (*Mucuna pruriens*/cowhage) 20mg.
2. *śilājīt* (*Asphaltum punjabianum*) 20mg.
3. *aśvagandhā* (*Withania somnifera*) 20mg.
4. *safed* (white) *mūslī* (*Asparagus adscendens*/*Chlorophytum arundinaceum*) 20mg.
5. *samūdraśoś/samūdra patra/vidhārā/vṛddhadārū* (Hawaiian baby woodrose/elephant creeper, *Argyreia speciosa*) 20mg.
6. *dālcīnī* (cinnamon) 20mg.
7. *sālam/sālab miśri* (*Orchis mascula/latifolia*) 20mg.
8. *gokhrū/gokśurā* (*Tribulus terrestris*) 20mg.
9. *kabāb cīnī* (tailed pepper/*Piper kubeba*) 20mg.
10. *vidārīkāṇḍ/vidārī/kudzu* (*Hedysarum tuberosum*/*Pueraria tuberosa*) 20mg.
11. *abhrak/abrak bhasm* (calcinated mica ash) 15mg.
12. *vaṅg/baṅg bhasm* (calcinated tin powder) 15mg.
13. *jāyphal* (*Myristica fragrans*/nutmeg) 15mg.
14. *jāvitrī* (fibrous coating of nutmeg) 15mg.
15. *akarkarā/ākārakarabha* (Spanish chamomile, *Anacyclus pyrethrum*) 10mg.
16. *lavaṅg/lauṅg* (clove, *Syzygium aromaticum*/*Caryophyllus aromaticus*) 10mg.
17. *raupya bhasm* (*candi*/calcinated silver ash) 6mg.
18. *kāṇt lauh bhasm* (calcinated iron ash plus other ingredients) 6mg.
19. *kesar* (saffron, *Crocus sativus*) 6mg.
20. *svarṇ bhasm* (calcinated gold powder) 1mg.
21. a concoction of *lāl mūslī/semal* (*Bombax malibaricum*), *śatāvarī, brahmi* (water hissop, *Bacopa monnieri*), *candan safed* (white sandalwood), *mulețhi* (liquorice), *pān* (betel nut, *Areca catechu*) juice, processed through the *bhāvanā* (infusion) using *ghana satva* (heating and drying of concentrated herbal extracts).[16]

2. Bherav Shakti (Ankur Herbals Pvt. Ltd, Delhi)

1. *aśvagandhā* (*Withania somnifera*) 100mg.
2. *ātmagupta/kauṃc/kevāñc/kapikacchu* (cowhage, *Mucuna pruriens*) 80mg.

---

[16] The term *satva* is usually used in reference to the sedimented, starchy, aqueous extract of *guḍūcī* (*Tinospora cordifolia*), while the term *ghana* (meaning 'dense') usually (but not always) refers to the solidified aqueous extract of the plant (Sharma and Prajapati 2017).

3. *kuclā/kucilā/kupīlu* (strychnine tree, *Strychnos nuxvomica*)[17] 20mg.
4. *amṛta* (corduroy orchid, *Eulophia campestris*) 30mg.
5. *gokhrū/gokṣurā* (*Tribulus terrestris*) 50mg.
6. *makardhvaj bhasm* (gold, mercury, sulphur: ratio 1:8:24) 80mg.
7. *kesar* (saffron, *Crocus sativus*) 20mg.
8. *svarṇ bhasm* (calcinated gold powder) 20mg.
9. *moti piṣṭi* [powdered] (pearl calcinated in rose water) 20mg.
10. *vaṅg/baṅg bhasm* (calcinated tin powder) 20mg.
11. *trivaṅg bhasm* (tin, lead, zinc: processed in aloe vera juice and turmeric) 20mg.
12. *akarkarā/ākārakarabha* (Spanish chamomile, *Anacyclus pyrethrum*) 20mg.
13. *candi bhasm* (*Argentum*) 20mg.
14. *svarṇ varaq* (gold foil, [L.] *aurum*) 10mg.

3. Commando (Dr Asma Herbals, Chhetakalan, Amritsar)

1. *kesar* (saffron, *Crocus sativus*) 20mg.
2. *safed* (white) *mūslī* (*Asparagus adscendens/Chlorophytum arundinaceum*) 50mg.
3. *akarkarā/ākārakarabha* (Spanish chamomile, *Anacyclus pyrethrum*) 50mg.
4. *sālampañjā/sālam miśri* (marsh orchid, *Orchis latifolia*) 50mg.
5. *aśvagandhā* (*Withania somnifera*) 50mg.
6. *tāl makhānā/kokilakṣa* (fox nut/prickly waterlily, *Asteracantha longifolia*) 20mg.
7. *śatāvarī/satvīrya* (*Asparagus racemosus*) 20mg.
8. *pīpal* (*Ficus religiosa*) 20mg.
9. *ātmagupta/kauṃc/kevāñc/kapicacchu* (cowhage, *Mucuna pruriens*) 20mg.
10. *jāyphal/jatiphal* (nutmeg, *Myristica fragrans*) 20mg.
11. *tulsī* (Indian basil, *Ocinum sanctum*) 20mg.
12. *gokhrū/gokṣurā* (*Tribulus terrestris*) 20mg.
13. *vidārīkāṇḍ/vidārī/kudzu* (*Hedysarum tuberosum/Pueraria tuberosa*) 20mg.
14. Ceylon cinnamon (*Cinnamomom zeylanicum*) 20mg.
15. cardamom (*Elattaria cardamomum*) 20mg.
16. myrobalan/*āṃvlā* (*Phyllanthus emblica/officinalis*) 50mg.
17. *kuclā/kucilā/kupīlu* (strychnine tree, *Strychnos nuxvomica*) 10mg.
18. *śilājīt* (*Asphaltum punjabianum*) 20mg.

4. Hi-Gra (Mittal Ayurved Sansthan, Meerut, U.P.)

1. *safed* (white) *mūslī* (*Chlorophytum arundinaceum/borivilianum*) 100mg.
2. *aśvagandhā* (*Withania somnifera*) 100mg.
3. *gokhrū/gokṣurā* (*Tribulus terrestris*) 100mg.
4. *śilājīt* (*Asphaltum punjabianum*) 50mg.
5. *śatāvarī/satvīrya* (*Asparagus racemosus*) 20mg.
6. *ātmagupta/kauṃc/kevāñc/kapikacchu* (cowhage, *Mucuna pruriens*)20mg.
7. marvel bronze/foxtail (*Amaranthus paniculatus*) 20mg.
8. *akarkarā/ākārakarabha* (Spanish chamomile, *Anacyclus pyrethrum*) 20mg.
9. *jāyphal/jatiphal/jāvitrī* (nutmeg, *Myristica fragrans*) 10mg.
10. *kuclā/kucilā/kupīlu* (strychnine tree, *Strychnos nuxvomica*) 10mg.

---

[17] Strychnine is highly poisonous (in relatively small doses) and is one of the substances more commonly added in Indiam to preparations of *bhāṅg* (edible cannabis). It is still occasionally combined with damiana as an aphrodisiac in Mexico and in south-western USA (Seldon 1979:191).

5. Japani M (Chaturbhuj Pharmaceutical Co., Hardwar, Uttarakhand)

1.	*śilājīt (Asphaltum punjabianum)*	25mg.
2.	*kesar* (saffron, *Crocus sativus*)	25mg.
3.	*aśvagandhā* (*Withania somnifera*) root	75mg.
4.	*ātmagupta/kauṃc/kevāñc/kapikacchu* (cowhage, *Mucuna pruriens*) seed	50mg.
5.	*akarkarā/ākārakarabha* (Spanish chamomile, *Anacyclus pyrethrum*) root	20mg.
6.	*śatāvarī/satvīrya* (*Asparagus racemosus*) root	50mg.
7.	*vaṅg/baṅg bhasm* (calcinated tin powder)	5mg.
8.	*chaglī/sāgaramekhalā* (beach morning glory, *Ipomoea biloba*) root (or ?) *vidhārā* (Hawaiian baby woodrose, *Argyreia speciosa*)	50mg.
9.	*safed* (white) *mūslī* (*Asparagus adscendens/Chlorophytum arundinaceum*)	50mg.
10.	*jāypatti/jāvitrī* (outer covering of mace/nutmeg)	10mg.
11.	*manmathābhra/manmatha ras* [juice]	30mg.

(The formula varies slightly for *manmathābhra* juice; below is an example)

kajjalī (black sulphide of mercury), *abhrak bhasm* (calcinated mica)-45.52 mg. (each of the above ingredients), *kapūr* (camphor)-7.27 mg., *vaṅg bhasm* (calcinated tin), *lauh bhasm* (calcinated iron)-22.75 mg. (each of the above ingredients), *tamra bhasm* (calcinated copper)-11.37 mg., *vidhārā* seed (*Argyreia speciosa*), *vidhārīkaṇḍ* (*Pueraria tuberosa*), *śatāvarī* (*Asparagus racemosus*), *tālmakhānā* (*Asteracantha longifolia*), *balā* (*Sida cordifolia*), *ati balā* (*Abutilon indicum*), *jāyphal* (*Myristica fragrans*), *jāvitrī* (*Myristica fragrans*), *lavaṅg* (*Syzygium aromaticum*), *bhāṅg* seed (*Cannabis sativa*), *sāl* tree (*Shorea robusta*), *ajvain* (*Trachyspermum ammi*)-7.27 mg. (each of the above ingredients).

12.	*abhrak/abrak bhasm* (calcinated mica ash)	10mg.
13.	*svarṇ bhasm* (calcinated gold powder)	7.5mg.
14.	*jāyphal/jatiphal/jāvitrī* (nutmeg, *Myristica fragrans*)	10mg.
15.	*ras sindūr* (sulphide [*gandhak*] of red mercury prepared in the juice of extract of banyan (*nyagrodha*)	10mg.
16.	*makardhvaj* (mercury sulphide [8 parts] and gold leaf [1 part])	10mg.
17.	*puṣpadhanvā ras*	30mg.

1. *ras sindūr* (sulphide of mercury), 2. *nāg bhasm* (calcinated lead, sulphur, lemon juice, manaśīlā [arsenic disulphide), 3. *lauh bhasm*, 4. *abhrak bhasm*, 5. *vaṅg bhasm*, 6. *dhatūrā*, 7. cannabis *(bhāṅg*), 8. *mulethī* (liquorice), 9. bark of *semal* (silk cotton tree, *Salmalia/Bombax malibaricum/ceiba*), 10. *nāgar bel* (betel/*pān*, *Areca catechu*).

18.	*kapūr* (camphor tree, *Cinnamomum camphora*)	5mg.

6. KS-Gold (Rasyanan Vidhya, Sonipat, Haryana)

1.	*svarṇ bhasm*	25mg.
2.	*raupya/raupyā/rūpya bhasm* (calcinated silver)	75mg.
3.	*mukta piṣṭī* (calcinated pearl infused in rosewater)	1mg.
4.	*vaṅg/baṅg bhasm* (calcinated tin powder)	20mg.
5.	*abhrak/abrak bhasm* (calcinated mica ash)	10mg.
6.	*lauh/lohā bhasm* (iron oxide)	20mg.
7.	*svarṇ mākśik bhasm* (chalcopyrite: copper, iron, sulphur)	8mg.
8.	*śilājīt* (*Asphaltum punjabianum*)	25mg.

9. cardamom (*Elattaria cardamomum*) 10mg.
10. *lavaṅg/lauṅg* (clove, *Syzygium aromaticum/Caryophyllus aromaticus*) 40mg.
11. *nāgkesar* (cobra saffron, *Mesua ferrea*) 40mg.
12. *aśvagandhā* (*Withania somnifera*) 40mg.
13. *ātmagupta/kauṃc/kevāñc/kapikacchu* (cowhage, *Mucuna pruriens*) 40mg.
14. *tejpatra* (Indian bay leaf, *Cinnamomum tamala*) 40mg.
15. *jāyphal/jatiphal/jāvitrī* (nutmeg, *Myristica fragrans*) 25mg.
16. *tāl makhānā/kokilakṣa* (fox nut/prickly waterlily, *Asteracantha longifolia/Hygrophila spinosa*) 40mg
17. cinnamon/*dālcīnī* 40mg.
18. *safed* (white) *mūslī* (*Asparagus adscendens/Chlorophytum arundinaceum*) 25mg.
19. *kapūr* (camphor tree, *Cinnamomum camphora*) 25mg.
20. *kuclā/kucilā/kupīlu* (strychnine tree, *Strychnos nuxvomica*) 25mg.

7. Madnānand Modak (Multani Pharmaceuticals, New Delhi)

1. *pārā/pārad* (mercury)
2. *gandhak* (yellow sulphur/brimstone)
3. *lauh/lohā bhasm* (iron oxide)
4. *kapūr* (camphor tree, *Cinnamomum camphora*)
5. *sendhā/saindhā* (pink rock salt)
6. [Nardostachis] *Jaṭāmāṅsī* (Indian spikenard)
7. *āṃvlā* (*Emblica officinalis/Phyllanthus emblica*)
8. cardamom/*ilāycī* (*Elattaria cardamomum*)
9. ginger powder/*soṇṭh*
10. chilli/*mirc*
11. *pippalī* (long pepper, *Piper longum*)
12. *jāyphal/jatiphal* (nutmeg, *Myristica fragrans*)
13. *jāypatti/jāvitrī* (outer covering of mace/nutmeg)
14. *tejpatra* (Indian bay leaf, *Cinnamomum tamala*)
15. *lauṅg* (clove, *Syzgium aromaticum*)
16. *kālā jīrā* (black caraway, *Nigella sativa*)
17. *jirā/śveta jirakā* (cumin seed, *Cuminum cyminum*)
18. *muleṭhī* (liquorice root, *Glycyrrhiza glabra*)
19. *vac/bac* (sweet flag, *Acorus calamus*)
20. *kūṭh* (Indian costus, *Saussurea costus*)
21. *haldī* (turmeric)
22. *devdāru/devdār* (Himalayan cedar, *Cedrus deodara*)
23. *hijjal/samūdra phal* (Indian oak/freshwater mangrove, *Barringtonia acutangula*)
24. *suhāgā* (borax)
25. *bhāraṅgī* (blue fountain bush, *Clerodendron serratum*)
26. ginger powder/*soṇṭh*
27. *nāgkesar* (cobra saffron, *Mesua ferrea*)
28. *kakraśṛṅgī/karkaṭśṛṅgī* (crab's claw, *Pistacia integerrima*)
29. *tālīspatar/tālīspatra* (Indian silver fir tree, *Abies webbiana*)
30. *munnakā* (raisin)
31. *citrak* (leadwort, *Plumbago indica/zeylanica/auriculta*), root bark
32. *dāṅtī mūl* (root)/*niśoth* (Indian jalap, *Ipomoea turpethum*)
33. *balā/bījband/khareṭī* (country mallow, *Sida cordifolia*)
34. *atibalā/kaṅghī* (Indian mallow, *Abutilon indicum*)

35. cinnamon/*dālcīnī*
36. coriander/*dhanīyā*
37. *gajapippalī* (oriental cashew, *Scindapsus officinalis*)
38. *śaṭī/gaṅdhpalāśī* (spiky ginger lily, *Hedychium spicatur*)
39. *sugaṅdhbālā/tagar* (Indian valerian/fragrant swamp mallow, *Pavonia odorata*)
40. *mustā/mothā* (nutgrass sedge, *Cyperus rotundus*)
41. *gandh prasāriṇī* (Chinese fever vine/stinkvine, *Paederia foetida*)
42. *vidārīkāṇḍ/vidārī/kudzu* (*Hedysarum tuberosum/Pueraria tuberosa*)
43. *śatāvarī/satvīrya* (*Asparagus racemosus*)
44. *āk/ārka/madār jaḍ* (root) (Sodom apple, *Calotropis gigantea*)
45. *ātmagupta/kauṃc/kevāñc/kapikacchu* (cowhage, *Mucuna pruriens*)
46. *gokhrū/gokṣurā* (*Tribulus terrestris*)
47. *samūdraśoś/samūdra patra/vidhārā/vṛddhadārū* (Hawaiiian baby woodrose/elephant creeper, *Argyreia speciosa*)
48. cannabis/*bhāṅg* seed

All above: 0.29%

49. *abhrak/abrak bhasm* (calcinated mica ash) 0.88%
50. *lāl mūslī/semal* (*Bombax malibaricum*) 3.82%
51. *ghṛt/ghī* (clarified butter) 5.88%
52. honey/*madhu* 5.88%

cinnamon, Indian bay leaf, cardamom, cobra saffron, camphor, pink rock salt, dried ginger, long pepper (each) 0.14%

53. sugar 58.82%
54. cannabis powder (fried in *ghī* and goat milk) 9.41%

8. Musli Pro (Deemark Healthcare, Delhi)

1. *safed* (white) *mūslī* (*Asparagus adscendens/Chlorophytum arundinaceum*) 160mg.
2. *aśvagandhā* (*Withania somnifera*) 75mg.
3. *ātmagupta/kauṃc/kevāñc/kapikacchu* (cowhage, *Mucuna pruriens*) 75mg.
4. *vidārīkāṇḍ/vidārī/kudzu* (*Hedysarum tuberosum/Pueraria tuberosa*) 25mg.
5. *gokhrū/gokṣurā* (*Tribulus terrestris*) 25mg.
6. *balā/bījband/kareṭī* (*Sida cordifolia*) 25mg.
7. *dhātupuṣṭī* (combination of *aśvagandhā, kappicacchū, śatāvarī, vidārī, abrakh bhasm, makardhvaj*) 15mg.
8. *mālāphal* (*rudrākṣ, Elaeocarpus ganitrus*) 10mg.
9. *pīchil/ban turaī* (ribbed gourd, *Luffa acutangula*) 10mg.
10. *mkāltī* (jasmine, *Jasminum grandiflorum*) 10mg.
11. *śatāvarī/satvīrya* (*Asparagus racemosus*) 20mg.
12. *tālpatri* (black mūslī, *Curculigo orchiodes*) 20mg.
13. *tāl makhānā/kokilakṣa* (fox nut/prickly waterlily, *Asteracantha longifolia*) 20mg.
14. *muñjāṭak* (*Eulophia campestris*) 10mg.
15. *uṭaṅgan* (acanthus, *Blepharis edulis*) 10mg.
16. *akarkarā/ākārakarabha* (Spanish chamomile, *Anacyclus pyrethrum*) 20mg.
17. *śilājīt* (*Asphaltum punjabianum*) 25mg.
18. *vaṅg/baṅg bhasm* (calcinated tin powder) 10mg.
19. *swvarṇ baṅg/vaṅg bhasm* (*vaṅg* [tin ash], *pārā* [mercury], *gandhak* [sulphur], *nausādar* [ammonium chloride], *kalmī śorā* [potassium nitrate/saltpeter]) 10mg.
20. *svarṇ sindūr* (*kumkum, Bixa orellana*) 10mg.
21. (*siddh*) *makardhvaj* (gold, mercury, sulphur: ratio 1:8:24) 10mg.

9. Musliwin Gold Plus (Ayursiddha Inc., Raja Khasa, Kangra, H.P./ Jeotrix Healthcare, Haldwani, Uttarakhand)

1. *tāl makhānā/kokilakṣa* (fox nut/prickly waterlily, *Asteracantha longifolia*) 60mg.
2. *suniṣṇak/caupatiyā* (water clove, *Marsilea minuta*) 32mg.
3. *ātmagupta/kauṃc/kevāṃc/kapicacchu* (*Mucuna pruriens*/cowhage) 42mg.
4. *aśvagandhā* (*Withania somnifera*) 64mg.
5. *latā kastūrī/muṣkdānā* (musk mallow, *Hibiscus abelmoschus/ Abelmoschus moschatus*) 32mg.
6. *semal* (silk cotton tree gum, *Salmalia malabarica*) 16mg.
7. *akarkarā/ākārakarabha* (Spanish chamomile, *Anacyclus pyrethrum*) 21mg.
8. *safed* (white) *mūslī* (*Asparagus adscendens*/*Chlorophytum arundinaceum*) 63mg.
9. *śatāvarī/satvīrya* (*Asparagus racemosus*) 40mg.
10. *śilājīt* (*Asphaltum punjabianum*) 37mg.
11. *kuclā/kucilā/kupīlu* (strychnine tree, *Strychnos nuxvomica*) 15mg.
12. *kesar* (saffron, *Crocus sativus*) 1mg.
13. *makardhvaj bhasm* (gold, mercury, sulphur: ratio 1:8:24) 7mg.
14. *trivaṅg bhasm* (tin, lead, zinc: processed in aloe vera juice and turmeric) 20mg.

10. Piyagra (Surjichem Herbs India, Delhi)

1. *lavaṅg/lauṅg* (clove, *Syzygium aromaticum/Caryophyllus aromaticus*) 50mg.
2. *jāyphal/jatiphal* (nutmeg, *Myristica fragrans*) 50mg.
3. *kuclā/kucilā/kupīlu* (strychnine tree, *Strychnos nuxvomica*) 30mg.
4. *kālī mirc* (black pepper, *Piper negrum*) 30mg.
5. *ātmagupta/kauṃc/kevāñc/kapicacchu* (cowhage, *Mucuna pruriens*) 70mg.
6. *akarkarā/ākārakarabha* (Spanish chamomile, *Anacyclus pyrethrum*) 100mg.
7. *śatāvarī/satvīrya* (*Asparagus racemosus*) 60mg.
8. *śilājīt* (*Asphaltum punjabianum*) 50mg.
9. *svarṇ baṅg/vaṅg bhasm* (*vaṅg* [tin ash], *pārā* [mercury], *gandhak* [sulphur], *nausādar* [ammonium chloride], *kalmī śorā* [potassium nitrate/saltpeter]) 25mg.
10. *lauh/lohā bhasm* (iron oxide) 25mg.

11. Retalvit Gold (Syncom Healthcare Ltd., Dehradun, Uttarakhand)

1. *ātmagupta/kauṃc/kevāñc/kapikacchu* (cowhage, *Mucuna pruriens*) 125mg.
2. *safed* (white) *mūslī* (*Asparagus adscendens*/*Chlorophytum arundinaceum*) 100mg.
3. *gokhrū/gokṣurā* (*Tribulus terrestris*) 75mg.
4. *aśvagandhā* (*Withania somnifera*) 50mg.
5. *śilājīt* (*Asphaltum punjabianum*) 50mg.
6. *jāyphal/jatiphal/jāvitrī* (nutmeg, *Myristica fragrans*) 25mg.
7. *kapūr* (camphor tree, *Cinnamomum camphora*) 5mg.
8. *yaśad/jasad bhasm* (calcinated zinc oxide prepared with aloe vera juice) 25mg.
9. *abhrak/abrak bhasm* (calcinated mica ash) 25mg.
10. *pravāl bhasm* (calcinated coral) 25mg.
11. *svarṇ bhasm* (calcinated gold powder) 0.5mg.

12. Santushti (Multani Phamaceuticals, New Delhi)

1. *śilājīt* (*Asphaltum punjabianum*) 50mg.

2. *ātmagupta/kauṃc/kevāñc/kapikacchu* (cowhage, *Mucuna pruriens*) 80mg.
3. *tāl makhānā/kokilakṣa* (fox nut/prickly waterlily, *Asteracantha longifolia*) 82mg.
4. *aśvagandhā* (*Withania somnifera*) 100mg.
5. *safed* (white) *mūslī* (*Chlorophytum arundinaceum/borivilianum*) 30mg.
6. *makardhvaj* (mercury sulphide [8 parts] and gold leaf [1 part]) 20mg.
7. *kuclā/kucilā/kupīlu* (strychnine tree, *Strychnos nuxvomica*) 10mg.
8. *jāypatti/jāvitrī* (outer covering of mace/nutmeg) 20mg.
9. *ahiphen*/opium (*Papaver somniferum*) 8mg.

13. Shahenshahi Alam (O. S. Ayurved Pvt., Ltd., Kundli, Haryana)

1. *śilājīt* (*Asphaltum punjabianum*) 20mg.
2. *kuclā/kucilā/kupīlu* (strychnine tree, *Strychnos nuxvomica*) 20mg.
3. *śatāvarī/satvīrya* (*Asparagus racemosus*) 20mg.
4. *jāyphal/jatiphal/jāvitrī* (nutmeg, *Myristica fragrans*) 10mg.
5. *tāl makhānā/kokilakṣa* (fox nut/prickly waterlily, *Asteracantha longifolia/Hygrophila spinosa*) 30mg.
6. *akarkarā/ākārakarabha* (Spanish chamomile, *Anacyclus pyrethrum*) 20mg.
7. *safed* (white) *mūslī* (*Asparagus adscendens/Chlorophytum arundinaceum*) 30mg.
8. *ātmagupta/kauṃc/kevāñc/kapikacchu* (cowhage, *Mucuna pruriens*) 30mg.
9. *kesar* (saffron, *Crocus sativus*) 1mg.
10. *lavaṅg/lauṅg* (clove, *Syzygium aromaticum/Caryophyllus aromaticus*) 10mg.
11. *sandras/candras* (*Vateria indica* tree (Indian copal, white dammar) 10mg.
12. *jāypatti/jāvitrī* (outer covering of mace/nutmeg) 20mg.
13. *aśvagandhā* (*Withania somnifera*) 30mg.
14. *sālam/sālab miśri* (*Orchis mascula/latifolia*) 30mg.
15. *kapūr* (camphor tree, *Cinnamomum camphora*) 20mg.
16. opium/*post dānā* (*Papaver somniferum*) 20mg.
17. Syrian rue/*isfand* (Peganum harmala) 15mg.
18. *mastagī rūmī*/mastic tree (*Pistacia lentiscus*) 15mg.
19. *vaṅg/baṅg bhasm* (calcinated tin powder) 30mg.
20. *ras sindūr* (sulphide [*gandhak*] of red mercury prepared in the juice of extract of banyan (*nyagrodha*) 30mg.
21. *moti piṣṭi* [powdered] (pearl calcinated in rose water) 5mg.
22. *abhrak/abrak bhasm* (calcinated mica ash) 30mg.
23. *bhilāvā* (marking nut tree, *Semecarpus anacardium*) 10mg.
24. *vatsnābh/amṛta/mīṭhā teliyā* (Indian aconite, *Aconitum ferox*) 14mg.
25. *lauh/lohā bhasm* (iron oxide) 30mg.

14. Shilapravang Special (Om Pharmaceuticals, Bangalore)

1. *śilājīt* (*Asphaltum punjabianum*) 40mg.
2. *pravāl bhasm* (calcinated coral) 20mg.
3. *vaṅg/baṅg bhasm* (calcinated tin powder) 20mg.
4. *suvarṇāmakśak bhasm* (calcinated copper and iron sulphide) 20mg.
5. *makardhvaj* (gold, mercury, sulphur: ratio 1:8:24) 10mg.
6. *suvarṇ bhasm* (calcinated gold) 1mg.
7. *mukta piṣṭī* (calcinated pearl infused in rosewater) 1mg.

8.	*guḍūcī satva* (water-extracted *giloy/amṛta, Tinospora cordifolia*) stem[18]	20mg.
9.	*aśvagandhā* (*Withania somnifera*)	60mg.
10.	*śatāvarī* (*Asparagus racemosus*)	15mg.
11.	*gokhrū/gokṣurā* (*Tribulus terrestris*)	15mg.
12.	*balā/bījband/khareṭī* (country mallow, *Sida cordifolia*) root	15mg.
13.	*āmalakī* (*āṃvlā, Emblica officinalis/Phyllanthus emblica*) fruit pericarp	10mg.
14.	*akarkarā/ākārakarabha* (Spanish chamomile, *Anacyclus pyrethrum*)	10mg.
15.	*jāyphal/jatiphal* (nutmeg, *Myristica fragrans*)	5mg.
16.	*kapūr* (camphor tree, *Cinnamomum camphora*)	5mg.
17.	*lāṭākastūrī/kastūrīdānā* (musk mallow, *Hibiscus abelmoscus*) seed	20mg.
18.	*ātmagupta/kauṃc/kevāñc/kapicacchu* (cowhage/*Mucuna puriens*)	90mg.

15. Stay-On (Maruti Herbal, Mumbai)

1.	*safed* (white) *mūslī* (*Asparagus adscendens/Chlorophytu arundinaceum*)	8%
2.	*jāyphal/jatiphal* (nutmeg, *Myristica fragrans*)	7%
3.	*aśvagandhā* (*Withania somnifera*) *satva*	6%
4.	*gokhrū/gokṣurā* (*Tribulus terrestris*) *satva*	5%
5.	*lavaṅg/lauṅg* (clove, *Syzygium aromaticum/Caryophyll aromaticus*)	4%
6.	*tāl makhānā/kokilakṣa* (fox nut/prickly waterlily, *Asteracant longifolia*)	5%
7.	*balā/bījband/khareṭī* (country mallow, *Sida cordifolia*)	4%
8.	*sālampañjā/sālam miśrī* (marsh orchid, *Dactylorhiza hatagirea/Orchis lfolia*)	4%
9.	*vidārīkāṇḍ/vidārī/kudzu* (*Hedysarum tuberosum/Puera tuberosa*)	1%
10.	*jāypatti/jāvitrī* (outer covering of mace/nutmeg)	6%
11.	*akarkarā/ākārakarabha* (Spanish chamomile, *Anacyc pyrethrum*)	6%
12.	*pīpal* (*Ficus religiosa*)	5%
13.	*ātmagupta/kauṃc/kevāñc/kapicacchu* (cowhage, *Muc pruriens*)	6%
14.	*śatāvarī/satvīrya* (*Asparagus racemosus*)	5%
15.	*uṭigan/uṭaṅgan/dūdhiyā coṭī* (Madras blepharis, *Blepharis edulis/maderaspatensis*	4%
16.	*śilājīt* (*Asphaltum punjabianum*)	3%
17.	*vaṅg/baṅg bhasm* (calcinated tin powder)	*2%*
18.	*(siddh makardhvaj* (gold, mercury, sulphur: ratio 1:8	2%
19.	ginseng	10%

16. Swarnvati (Tulison Pharma, Delhi)

1.	(*siddh*) *makardhvaj* (gold, mercury, sulphur: ratio 1)	25mg.
2.	*kapūr* (camphor tree, *Cinnamomum camphora*)	10mg.
3.	*śatāvarī/satvīrya* (*Asparagus racemosus*) *cūrṇ* (pov	125mg.
4.	*mājū phal*/fruit (gall oak, *Querus infectoria*)	20mg.
5.	*guḍūcī satva* (water-extracted *giloy/amṛta, Tinosp rdifolia*) stem	20mg.
6.	*śilājīt* (*Asphaltum punjabianum*)	25mg.
7.	*kuclā/kucilā/kupīlu* (strychnine tree, *Strychnos nux a*)	25mg.

[18] The stem of *guḍūcī* is used (Shilpa and Chaudhary 2013).

17. Tentex Forte (Himalaya Herbal Healthcare, Bengaluru)

1. *lāṭākastūrī/kastūrīdānā* (musk mallow, *Hibiscus abelmoschus*) seed 10mg.
2. *aśvagandhā* (*Withania somnifera*) 65mg.
3. *samūdraśoś/samūdra patra/vidhārā/vṛddhadārū*
   (Hawaiian baby woodrose/elephant creeper, *Argyreia speciosa*) 32mg.
4. *ātmagupta/kauṃc/kevāñc/kapikacchu* (cowhage, *Mucuna pruriens*) 32mg.
5. *trivaṅg bhasm* (tin lead, zinc: processed in aloe vera juice and turmeric) 32mg.
6. *śilājīt* (*Asphaltum punjabianum*) 32mg.
7. *kumkum* (*Bixa orelana*) 25mg.
8. *kuclā/kucilā/kupīlu* (strychnine tree, *Strychnos nuxvomica*) 16mg.
9. *makardhvaj* (mercry sulphide [8 parts] and gold leaf [1 part]) 16mg.
10. *balā/bījband/kharē* (country mallow, *Sida cordifolia*) root 16mg.
11. *lāl mūslī/semal/śālmli/kapok*
    (silk cotton tree, *Samalia/Bombax malibaricum/ceiba*) 16mg.
12. black pepper/*kālī mr* (*Piper nigrum*) 5mg.

Processed in:

balā/bījband/khare (country mallow, *Sida cordifolia*)
śatāvarī/satvīrya (*paragus racemosus*)
vidārīkāṇḍ/vidārī/kzu (*Hedysarum tuberosum/Pueraria tuberosa*)
nagavalli (betel leapān], *Areca catechu*), *Piper betel*)
aśvagandhā (*Witha somnifera*)
gokhrū/gokṣurā (*Tulus terrestris*)
guḍūcī (*giloy/amṛt* *inospora cordifolia*)
samūdraśoś/samūdpatra/vidhārā/vṛddhadārū
(Hawaiian by woodrose/elephant creeper, *Argyreia speciosa*)
khādir/khair (cutche, *Acacia catechu*)
daśamūla (ten roots

18. Tentex Royal (HimalaHerbal Healthcare, Bengaluru)

1. *tāl makhānā/kokila* (fox nut/prickly waterlily, *Asteracantha longifolia*) 145mg.
2. *bādām/vātāda* (alm, *Prunus amygdalus*) 126mg.
3. **suniṣaṇṇaka* (*Man quadrifolia*) *Blepharis edulis**? 115mg.
4. *kesar* (saffron, *Croatius*) 14mg.
5. *gokhrū/gokṣurā* (*Tus terrestris*) 100mg.
   processed in: *kālīk*) *mūslī* (*Curculigo orchiodes*) and
   nag (betel leaf [*pān*], *Areca catechu*), *Piper betel*

19. Titanic Extra Time (Sdia Pharmacy, Orai, U.P.)

1. *aśvagandhā* (*Withomnifera*) 60mg.
2. *śatāvarī/satvīrya* (*agus racemosus*) 60mg.
3. *ātmagupta/kauṃc/k/kapicacchu* (cowhage, *Mucuna pruriens*) 90mg.
4. *safed* (white) *mūslīragus adscendens/Chlorophytum arundinaceum*) 60mg.
5. *harītaki/harrā* (blarobalan tree/*Terminalia chebula*) 60mg.
6. *jāyphal/jatiphal* (n *Myristica fragrans*) 50mg.
7. ginger (*Zingiber ofe*) 60mg.
8. *makardhvaj bhasm* mercury, sulphur: ratio 1:8:24) 5mg.
9. *trivaṅg bhasm* (tin,zinc: processed in aloe vera juice and turmeric) 5mg.

10. opium (*Papaver somniferum*) 30mg.
11. *śilājīt* (*Asphaltum punjabianum*) 50mg.

20. Titanic K2 (Sun Laboratories, Orai, U.P.)

1. *aśvagandhā* (*Withania somnifera*) 240mg.
2. *arjun* tree (*Terminalia arjuna*) 100mg.
3. *safed* (white) *mūslī* (*Asparagus adscendens/Chlorophytum arundinaceum*) 100mg.
4. *kesar* (saffron, *Crocus sativus*) 50mg.
5. *abhrak/abrak bhasm* (calcinated mica ash) 50mg.
6. *makardhvaj bhasm* (gold, mercury, sulphur: ratio 1:8:24) 50mg.
7. *akarkarā/ākārakarabha* (Spanish chamomile, *Anacyclus pyrethrum*) 100mg.
8. *svarṇ bhasm* (calcinated gold powder) 10mg.

21. Veeon Feel 21 (Veeon Healthcare, New Delhi)

1. *kesar* (saffron, *Crocus sativus*) 2.5mg.
2. *candi bhasm* (*Argentum*) 2.5mg.
3. *lauh/lohā bhasm* (iron oxide) 20mg.
4. *śilājīt* (*Asphaltum punjabianum*) 25mg.
5. *aśvagandhā* (*Withania somnifera*) 25mg.
6. *akarkarā/ākārakarabha* (Spanish chamomile, *Anacyclus pyrethrum*) 25mg.
7. *śatāvarī/satvīrya* (*Asparagus racemosus*) 25mg.
8. *moti* (black-lip pearl, *Pinctada margaritifera*) 25mg.
9. *āṃvlā* (*Emblica officinalis/Phyllanthus emblica*) 25mg.
10. *gokhrū/gokṣurā* (*Tribulus terrestris*) 25mg.
11. *babūl* (*Acacia arabica*) 25mg.
12. *jāyphal/jatiphal/jāvitrī* (nutmeg, *Myristica fragrans*) 25mg.
13. *jyotiśmiti* seed (*Celastruspaniculatus*) 25mg.
14. *giloy* (*Tinospora cordifolia*) 25mg.
15. *lavaṅg/lauṅg* (clove, *Syzygium aromaticum/Caryophyllus aromaticus*) 25mg.
16. *mulethi* root (liquorice, *Azadirachta indica*) 25mg.
17. *tulsī* seed (*Ocinum sanctum*) 25mg.
18. *vidārīkāṇḍ/vidārī/kudzu* (*Hedysarum tuberosum/Pueraria tuberosa*) 40mg.
19. *tāl makhānā/kokilakṣa* (fox nut/prickly waterlily, *Asteracantha longifolia*) 40mg.
20. *safed* (white) *mūslī* (*Asparagus adscendens/Chlorophytum arundinaceum*) 40mg.

22. Vega Power (Vee Excel Drugs and Pharmaceuticals, Delhi)

1. *aśvagandhā* (*Withania somnifera*) root 90mg.
2. *uttanjan* (*Blefaris edulis*) seed 50mg.
3. *ātmagupta/kauṃc/kevāñc/kapicacchu* (cowhage/*Mucuna pruriens*) 50mg.
4. *śilājīt* (*Asphaltum punjabianum*) 30mg.
5. *safed* (white) *mūslī* (*Asparagus adscendens/Chlorophytum arundinaceum*) 30mg.
6. *lāl mūslī/semal/śālmali/kapok*
(silk cotton tree, *Salmalia/Bombax malibaricum/ceiba*) 25mg.
7. *kapok* flower 30mg.
8. myrobalan/*āṃvlā* (*Phyllanthus emblica/officinalis*) 25mg.
9. ginger (*Zingiber officinale*) 20mg.
10. *jāyphal/jatiphal* (nutmeg, *Myristica fragrans*) 10mg.

11. *kesar* (saffron, *Crocus sativus*) 5mg.

23. Vitomanhills (Isha Agro Developers Pvt. Ltd., Mumbai)

1. *aśvagandhā* (*Withania somnifera*) 117mg.
2. *safed* (white) *mūslī* (*Asparagus adscendens/Chlorophytum arundinaceum*) 39mg.
3. *śatāvarī/satvīrya* (*Asparagus racemosus*) 39mg.
4. *ātmagupta/kauṃc/kevāñc/kapikacchu* (cowhage, *Mucuna pruriens*) 117mg.
5. *vidārīkāṇḍ/vidārī/kudzu* (*Hedysarum tuberosum/Pueraria tuberosa*) 39mg.
6. *tāl makhānā/kokilakṣa* (fox nut/prickly waterlily, *Asteracantha longifolia/Hygrophila spinosa*) 39mg.
7. *jāyphal/jatiphal/jāvitrī* (nutmeg, *Myristica fragrans*) 10mg.
8. *phaseolus mungo/vigna mungo/māṣa* (black lentils, *uḍad dāl*) 39mg.
9. *akarkarā/ākārakarabha* (Spanish chamomile, *Anacyclus pyrethrum*) 19mg.

24. Why Not 12 (Zee Laboratories, New Delhi)

1. L-arginine (amino acid) 50mg.
2. ginkgo biloba extract 50mg.
3. ginseng 42.5mg.
4. *aqśvagandhā* (*Withania somnifera*) 50mg.
5. maca (*Lepidium myenii*) 50mg.
6. yohimbe 50mg.
7. vitamin E 20i.u.
8. vitamin B3 40mg.
9. L-lysine (amino acid) 40mg.
10. beta-alanine (amino acid) 50mg.
11. creatine 40mg.

25. Zandu Vigorex (Emami Limited, Kolkata)

1. *aśvagandhā* (*Withania somnifera*)
2. *śatāvarī/satvīrya* (*Asparagus racemosus*)
3. *ātmagupta/kauṃc/kevāñc/kapikacchu* (cowhage, *Mucuna pruriens*)
4. *safed* (white) *mūslī* (*Chlorophytum arundinaceum/borivilianum*)
5. *śilājīt* (*Asphaltum punjabianum*)
6. *gokhrū/gokṣurā* (*Tribulus terrestris*)
7. *yaśad/jasad bhasm* (calcinated zinc oxide prepared with aloe vera juice)

26. Zealot Gold (Link Pharma, Kundli, Haryana)

1. *safed* (white) *mūslī* (*Asparagus adscendens/Chlorophytum arundinaceum*) 200mg.
2. ginkgo biloba extract 45mg.
3. *gotu kola* (*Cantella asiatica/Mandooka pami*) 60mg.
4. *ātmagupta/kauṃc/kevāñc/kapikacchu* (cowhage, *Mucuna pruriens*) 50mg.
5. *gokhrū/gokṣurā* (*Tribulus terrestris*) 30mg.
6. *aśvagandhā* (*Withania somnifera*) ?
7. *śilājīt* (*Asphaltum punjabianum*) 50mg

12.2 FOR MEN and WOMEN

27. Energic-31 (Ayurved Vikras Sansthan, Kashipur, Uttarakhand)

1.	*śilājīt* (*Asphaltum punjabianum*)	450mg.
2.	*śaṅkh* (seashell) *bhasm*	10mg.
3.	*trivaṅg bhasm* (tin, lead, zinc: processed in aloe vera juice and turmeric)	30mg.
4.	*kuclā/kucilā/kupīlu* (strychnine tree, *Strychnos nuxvomica*)	50mg.
5.	*kukkuṭāṇḍtvak* (eggshell) *bhasm*	20mg.
6.	*muktaśukti bhasm/mukta piṣṭī* (calcinated pearl infused in rosewater)	10mg.
7.	*svarṇ mākśik bhasm* (chalcopyrite: copper, iron, sulphur)	10mg.
8.	*śatāvarī/satvīrya* (*Asparagus racemosus*)	20mg.
9.	*ātmagupta/kauṃc/kevāñc/kapikacchu* (cowhage, *Mucuna pruriens*) seed	10mg.
10.	*aśvagandhā* (*Withania somnifera*)	20mg.
11.	cinnamon/*dālcīnī*	10mg.
12.	*nāgkesar* (cobra saffron, *Mesua ferrea*)	10mg.
13.	*gokhrū/gokṣurā* (*Tribulus terrestris*)	10mg.
14.	dried ginger/*soṇṭh*	10mg.
15.	*lauh/lohā bhasm* (iron oxide)	10mg.
16.	*lodh paṭhānī* (lodh tree, *Symplocos racemosa*)	10mg.
17.	cardamom/*ilāycī* (*Elattaria cardamomum*)	10mg.
18.	*jāypatti/jāvitrī* (outer covering of mace/nutmeg)	10mg.
19.	*mīṭhā/śirīn surañjan* (Autumn crocus, *Colchicum luteum*)	10mg.
20.	*samūdraśoś/samūdra patra/vidhārā/vṛddhadārū* (Hawaiian baby woodrose/elephant creeper, *Argyreia speciosa*)	10mg.
21.	*jāyphal/jatiphal/jāvitrī* (nutmeg, *Myristica fragrans*)	10mg.
22.	*safed* (white) *mūslī* (*Chlorophytum arundinaceum/borivilianum*)	10mg.
23.	*samūdraśoś* (Hawaiian baby woodrose/elephant creeper, *Argyreia speciosa*)	10mg.
24.	*lavaṅg/lauṅg* (clove, *Syzygium aromaticum/Caryophyllus aromaticus*)	10mg.
25.	*babūl goṇd* (thorny acacia tree, *Acacia arabica/nilotica/Vachellia nilotica*)	10mg.
26.	*tāl makhānā/kokilakṣa* (fox nut/prickly waterlily, *Asteracantha longifolia*)	10mg.
27.	*chhoṭī pīpal* (short pepper, *Piper retrofactum*)	10mg.
28.	*kālī mirc* (black pepper, *Piper negrum*)	10mg.
29.	*safed candan* (white sandalwood, *Pterocarpus santalinus*)	10mg.
30.	*akarkarā/ākārakarabha* (Spanish chamomile, *Anacyclus pyrethrum*)	10mg.
31.	*kaṅkol mirc* (*Piper cubeba*)	

28. Semax (Ayurvedic Research Foundation, Bhahadurgarh, Haryana)

1.	*guggul*/Indian bdelium/*mukul* myrrh tree (*Commiphora mukul*)	20mg.
2.	*śilājīt* (*Asphaltum punjabianum*)	10mg.
3.	*tāl makhānā/kokilakṣa* (fox nut/prickly waterlily, *Asteracantha longifolia*)	20mg.
4.	*śatāvarī/satvīrya* (*Asparagus racemosus*)	20mg.
5.	*safed* (white) *mūslī* (*Asparagus adscendens/Chlorophytum arundinaceum*)	20mg.
6.	Indian snakeroot/*sarpgandhā* (*Rauwolfia serpentina*)	20mg.
7.	*jīvantī* orchid (*Desmotrichum fimbriatum*)	10mg.
8.	*mustā/mothā* (nutgrass sedge, *Cyperus rotundus*)	20mg.
9.	*atīs* (Greenish Himalayan monkshood/*Aconitum heterophyllum*)	10mg.
10.	*dhaniyā*/coriander (*Coriandrum sativum*)	10mg.
11.	*jaṭāmānsī* (musk root/Indian spikenard/*Nardostachys jatamansi*)	10mg.

12. *gokhrū/gokṣurā* (*Tribulus terrestris*) 10mg.
13. *svarṇ baṅg/vaṅg bhasm* (*vaṅg* [tin ash], *pārā* [mercury], *gandhak* [sulphur], *nausādar* [ammonium chloride], *kalmī śorā* [potassium nitrate/saltpeter]) 10mg.
14. *abrakh/abrak bhasm* (calcinated mica) 5mg.
15. *lauh/lohā bhasm* (iron oxide) 5mg.

The ingredients are processed in decoctions of:

16. *guḍūcī* (*giloy/amṛta, Tinospora cordifolia*)
17. *haldī* (turmeric, *Curcuma longa*)
18. *dāruhaldī/dāruharidrā* (tree turmeric/*Berberis aristata*)
19. *circitā* (goji berry, *Lycium barbarum*)
20. *citrak* (Ceylon leadwort, *Plumbago zeylanica*)
21. *triphalā* ('three fruits'):
    1. *āmalakī/āṃvlā, Emblica officinalis/Phyllanthus emblica*)
    2. *bibhītaki/vibhītaki*/belleric myrobalan/*baheḍ/Terminalia bellirica*
    3. *harītaki*/black myrobalan tree/*Terminalia chebula*)
22. *devdāru* (Himalayan cedar, *Cedrus deodara*)
23. Indian snakeroot/*sarpgandhā* (*Rauwolfia serpentina*)
24. *nisoth/niśoth* (St. Thomas lidpod, *Operculina turpethum*)
25. *ajvāyan/ajvāin khurāsānī/khursānī* (black henbane, *Hyoscyamus niger*)

12.3 FOR WOMEN

29. Japani-F (Chaturbhuj Pharmaceutical Co., Hardwar, Uttarakhand)

1. *śilājīt* (*Asphaltum punjabianum*) 60mg.
2. *kesar* (saffron, *Crocus sativus*) stigma 2.5mg.
3. *aśvagandhā* (*Withania somnifera*) 50mg.
4. *ātmagupta/kauṃc/kevāñc/kapikacchu* (cowhage, *Mucuna pruriens*) 60mg.
5. *akarkarā/ākārakarabha* (Spanish chamomile, *Anacyclus pyrethrum*) 25mg.
6. *gokhrū/gokṣurā* (*Tribulus terrestris*) 25mg.
7. *vaṅg/baṅg bhasm* (calcinated tin powder) 5mg.
8. *mājūphal* (dyer's oak/oak gall, *Quercus infectoria*) fruit 10mg.
9. *safed* (white) *mūslī* (*Chlorophytum arundinaceum/borivilianum*) root 40mg.
10. ashok tree (*Saraka indica*) bark 80mg.
11. *lodhra/lodh* tree (*Symplocos racemosa*) bark 25mg.
12. *āmalakī rasāyan* (*āṃvlā, Emblica officinalis/Phyllanthus emblica*; honey, ghī, *pīpal* [long pepper, *Piper longum*] powder; sugar; *āṃvlā* juice) 50mg.
13. *pradarāntak lauh*
    lauh (iron) *bhasm*, *tamra* (copper) *bhasm*, *hartāl bhasm* (orpiment/arsenic trisulphide), *vaṅg/baṅg bhasm* (calcinated tin powder), *abhrak bhasm*, *varāṭikā bhasm* (cowry shell, *Cyprea moneta*), dried ginger/*soṇṭh*, long pepper (*Piper longum*), *harītaki/harrā* (black myrobalan tree/*Terminalia chebula*), *vibhītaki* (Belleric myrobalan/*baheḍā/Terminalia bellirica*), *āmalalakī* (*Emblica officinalis*), castor oil plant/*eraṇḍa* (*Ricinus communis*), *vidāṅg* (false black pepper, *Embelia ribes*), *sendhā/saindhā* (pink rock salt), sea salt/*samūdra lavāṇā*, *viḍa/biḍa lavāṇā* (black/ammonium salt), *sauvarcalā/sauvarkalā* (sochal black salt), *aubhida lavāṇā* (*reha* salt), Java long pepper/*cavya* (*Piper chaba*), *śaṅkh* (seashell) *bhasm*, *vac/bac* (sweet flag, *Acorus calamus*), Juniper tree/*hapuṣā* (*Juniperus communis*), *kuṣṭha* (snow lotus, *Sassurea lappa*), *śaṭī/kapūr kacrī* (spiked ginger lily, *Hedychium spicatum*), *pāṭhā* (Indian moonseed, *Cyclea peltata*), *devdāru/devdār* (Himalayan cedar, *Cedrus deodara*), cardamom (*Elattaria cardamomum*), *samūdraśoś/samūdra patra/vidhārā/*

vṛddhadārū (Hawaiian baby woodrose/elephant creeper, *Argyreia speciosa*)

14. *trivaṅg bhasm* (tin, lead, zinc: processed in aloe vera juice and turmeric) 12.5mg.
15. *puṣpadhanvā ras* 50mg.

1. *ras sindūr* (sulphide of mercury), 2. *nāg bhasm* (calcinated lead, sulphur, lemon juice, *manaśīlā* [arsenic disulphide), 3. *lauh bhasm*, 4. *abhrak bhasm*, 5. *vaṅg bhasm*, 6. *dhatūrā*, 7. cannabis *(bhāṅg)*, 8. *muleṭhī* (liquorice), 9. bark of *semal* (silk cotton tree, *Salmalia/Bombax malibaricum/ceiba*), 10. *nāgar bel* (betel/*pān, Areca catechu*).

30. Labisa Pumila Complex (BGSG, Kuala Lumpur)

1. *Labisia pumila/kacip Fatima* 154mg.
2. turmeric (*Curcumae longae rhizoma*) 49mg.
3. ginko biloba 49mg.
4. *Carthami flos* (safflower/*Carthamus tinctorius*) 24.5mg.
5. *Herba epimdii/yinyanghuo* 24.5mg.
6. *Cistanches herba/roucongrong* (desert ginseng, *Cistanche deserticola*) 24.5mg.
7. *Astragalus membranaceus/huanghuāhuangqi* (Mongolian milkvetch) 24.5mg.

31. Vigoroyal-1 (Maharishi Ayurveda Products Pvt. Ltd., New Delhi)

1. *aśok* (Ashoka tree, *Saraca indica*) 25mg.
2. *lodh paṭhānī* (lodh tree, *Symplocos racemosa*) 25mg.
3. *lajjālū* (shame plant, *Mimosa pudica*) 10mg.
4. *pataṅg kāṣṭ* (sappanwood tree, *Caesalpinia sappan*) 75mg.
5. *balā/bījband/khareṭī* (country mallow, *Sida cordifolia*) 135mg.
6. *gokhrū/gokṣurā* (*Tribulus terrestris*) 50mg.
7. *mustā/mothā/nāgar mothā* (nutgrass sedge, *Cyperus rotundus*) 10mg.
8. *nīlophar/nīlotpal* (blue lotus, *Nymphaea nouchali*) 10mg.
9. white sandalwood/*safed candan* (*Santalum album*) 10mg.
10. ginger/*soṇṭh* (*Zingiber officinale*) 10mg.
11. *pippalī* (long pepper, *Piper longum*) 10mg.
12. *aśvagandhā* (*Withania somnifera*) 70mg.
13. *vidārīkāṇḍ/[kṣīr/ vidārī/kudzu*
    (*Ipomoea digita/Hedysarum tuberosum/Pueraria tuberosa*) 40mg.
14. *samūdraśoś/samūdra patra/vidhārā/vṛddhadārū*
    (Hawaiian baby woodrose/elephant creeper, *Argyreia speciosa*) 15mg.
15. *śatāvarī/satvīrya* (*Asparagus racemosus*) 165mg.
16. *nāgkesar* (cobra saffron, *Mesua ferrea*) 5mg.
17. cumin/caraway/*jīrā* (*Cuminum cyminum*) 5mg.
18. black cumin/black caraway/*kālā jīrā* (*Carum bulbocastanum*) 5mg.
19. *śilājīt* (*Asphaltum punjabianum*) 50mg.
20. red coral powder/*pravāl bhasm* (*Corallium rubrum*) 10mg.
21. *abhrak/abrak bhasm* (calcinated mica ash) 5mg.
22. *dhātri lauh bhasm* (*āṃvlā*, liquorice [*muleṭhī*], iron oxide) 10mg.

32. 2 Much (Pripha Pharma/Prince Pharma, Ludhiana, Punjab)

1. *kuclā/kucilā/kupīlu* (strychnine tree, *Strychnos nuxvomica*) 60mg.
2. *ātmagupta/kauṃc/kevāṃc/kapicacchu* (*Mucuna pruriens*/cowhage) 50mg.
3. *śatāvarī/satvīrya* (*Asparagus racemosus*) 40mg.

4. *jyotiśmatī/māl kāṅgnī* (intellect/climbing staff tree, *Celastrus peniculatus*) 40mg.
5. *balā/bījband/khareṭī* (country mallow, *Sida cordifolia*) 40mg.
6. *aśvagandhā* (*Withania somnifera*) 40mg.
7. *safed* (white) *mūslī* (*Chlorophytum arundinaceum/borivilianum*) 25mg.
8. liquorice/*mulеṭhī* (*Glycyrrhiza glabra*) root 25mg.
9. datura (*Datura metel*) seed 20mg.
10. Ceylon cinnamon (*Cinnamomum zeylanicum*) 20mg.
11. *lavaṅg/lauṅg* (clove, *Syzygium aromaticum/Caryophyllus aromaticus*) 20mg.
12. *sālampañjā/sālam miśri* (marsh orchid, *Orchis latifolia*) 15mg.
13. *akarkarā/ākārakarabha* (Spanish chamomile, *Anacyclus pyrethrum*) 10mg.
14. *ajvāyan/ajvāin/khurāsānī/khursānī* (black henbane, *Hyoscyamus niger*) 10mg.
15. *kālī* (black) *mūslī* (*Curculigo orchiodes*) 20mg.
16. *uṭigan/uṭaṅgan/dūdhiyā coṭī* (Madras blepharis, *Blepharis edulis/maderaspatensis*) 20mg.
17. *safed* (white) behman (*Centarea behen*) 18mg.
18. *regmāhī* (common keeled skink/sand lizard, *Mabuya/Eutropis carinata*) 1mg.
19. *jūṇḍ beḍāsṭār* (*Castoreum*) [from glands near the reproductive organs of the castor beaver] 1mg.
20. *kesar* (saffron, *Crocus sativus*) 5mg.

Part 4
Non-Indian Aphrodisiac Formulas

13.1 *Non-Indian aphrodisiac formulas (untested)*

There are now very many male and female aphrodisiac formulas on sale globally. Several surveys of products in the USA of these products for men[19] reveal that many of these combination formulas use plants that feature in Āyurvedic prescriptions, but also other ingredients that are not South Asian. The reviews were sponsored, so they do not necessarily represent the most popular or effective products available. The author has not yet taken any of these formulas. The ingredients of the aphrodisiac supplement formulas as provided in the articles referenced above and on other websites are not entirely consistent, so the formulas listed below may possibly not be completely accurate.

33. Male Extra

1. ellagic acid (from pomegranate), which is in many fruits and vegetables; the highest concentrations are in yellow raspberries and pomegranate
2. L-arginine HCL (amino acid)
3. methyl sulfonyl methane
4. L-methionine (amino acid)
5. coryceps
6. zinc
7. niacin

34. Max Performer

1. horny goat weed
2. maca root extract
3. Korean red ginseng
4. cordyceps
5. bioperine (extract of black pepper)
6. selenium
7. zinc
8. niacin
9. pantothenic acid (Vitamin B5)
10. pyridoxine HCL (Vitamin B6)
11. iron
12. riboflavin
13. cyanocabolamin (Vitamin B12)

35. Performer 8

1. *muira puama*
2. *aśvagandhā*
3. maca root extract
4. panax ginseng

---

[19] Cordin (2021); *Houston Press* (2021); *Men's Journal* (2021).

5. ferrous bisglycinate
6. pine bark extract
7. glucuronolactone
8. horny goat weed (barrenwort)
9. grape seed extract

36. ProSolution Plus

1. *Tribulus terrestris*
2. *Withania somnifera* (*aśvagandhā*)
3. *Asparagus adscendens* (white *mūslī*)
4. *Mucuna pruriens* (*ātmagupta*)
5. *Asteracantha longifolia* (prickly waterlily)
6. *Curculigo orchiodes* (*black mūslī*)
7. solidilin
8. Korean ginseng
9. *Butea superba*
10. *momordica* (Bitter melon, *Momordica charantia*)
11. *āmlā* (*āṃvlā*)
12. Arjuna tree (*Terminalia arjuna*)
13. cordyceps
14. zinc
15. reishi mushroom (*lingzhi*)
16. drilizen
17. bladderwrack (Sea oak, *Ficus vesiculosus*)
18. apigenin (4, 5, 7 trihydroxy flavone)

37. Semenax

1. maca
2. *Muira puama*
3. pumpkin seed (for zinc)
4. zinc oxide
5. *Epimedium sagittatum* (horny goat weed)
6. *Avena sativa* (oats) extract
7. sarsaparilla
8. Swedish flower pollen
9. cranberry extract
10. panax ginseng
11. saw palmetto
12. gingko biloba

38. TestoPrime

1. D-Aspartic acid (amino acid)
2. ginseng
3. *aśvagandhā*
4. fenugreek
5. green tea extract with 70% catechins (also found in cocoa and grapes)
6. pomegranate

7. vitamin D
8. vitamin B5
9. Vitamin B6
10. garlic extract
11. black pepper extract
12. zinc

39. Viasil

1. *Epimedium brevicornum* (horny goat weed)
2. zinc
3. *Citrus sinensis* (orange blossom)
4. ginko biloba
5. *Tribulus terrestris*
6. panax ginseng
7. pomegranate

40. VigRXPlus

1. *Muira puama* bark extract
2. *epimedium* leaf extract (horny goat weed)
3. Asian red ginseng
4. damiana flower
5. hawthorn berry
6. *catuaba* bark extract
7. saw palmetto
8. saw palmetto berry
9. gingko biloba
10. bioperine (extract of black pepper)
11. Korean red ginseng
12. damiana leaf
13. *Tribulus terrestris*
14. cuscuta (dodder) seed extract

41. Volume Pills

1. *xi lan rou gui* (cinnamon bark, *Cinnamomum zylanicum,*)
2. *hong hua* [*fen*] (safflower flower, *Cathamus tinctorius/Flos cathami*)
3. *kū guā* (bitter melon, *Momordica charantia*)
4. 4, 5, 7 trihydroxy flavone (apigenin)
5. *Emblica officinalis* (*āṃvlā*)
6. *san guo mu* (chebulic myrobalan, *Terminalia chebula*)
7. *dong chong xia cao* (*Coryceps militaris*)
8. *xian mao* (black *mūslī*)
9. zinc gluconate
10. solidilin (contains L-Dopa) [this seems to be a novel chemical]
11. drilizen (contains protodioscin) [this seems to be a novel chemical]
12. *tian men dong* (asparagus root, *Radix asparagi*)
13. reishi mushroom (*lingzhi*)
14. *Tribulus terrestris*

13.2 *Non-Indian aphrodisiac formulas (tested)*

42. Hard Rod Plus

1.	*Tongkat ali* (*Eurycoma longifolia*)	75mg.
2.	*suo yang* (*Cynomorium songaricum*)	75mg.
3.	horny goat weed (*Epimedium acumiunatum*)	50mg.
4.	*gokhrū/gokṣurā* (*Tribulus terrestris*)	50mg.
5.	*Butea superba*	50mg.
6.	*Smilax myosotiflora*	25mg.
7.	*Diascoria bulbifera*	350mg.

43. Libidus (Bio-Gulf International Corporation)

1.	*Tongkat ali* (*Eurycoma longifolia*)	154mg.
2.	*Carthami flos* (safflower/*Carthamus tinctorius*)	24.5mg.
3.	turmeric (*Curcumae longae rhizoma*)	49mg.
4.	ginko biloba/Maidenhair tree	49mg.
5.	*Herba epimdii/yinyanghuo* (Horny goat weed)	24.5mg.
6.	*Cistanches herba/roucongrong* (desert ginseng, *Cistanche deserticola*)	24.5mg.
7.	*Astragalus membranaceus/huanghuāhuangqi* (Mongolian milkvetch)	24.5mg.

PART 5
The MAOI effect

14.1 *Monoamine oxidases (MAO)*

Monoamine oxidases (MAOs) are a family of enzymes[20] found in the brain, gut, liver and other tissues; they catalyse the oxidation and inactivation of monoamine neurotransmitters, including serotonin, noradrenaline and dopamine, which are important in the regulation of mood.

There are two kinds of MAO: MAO-A and MAO-B. MAO-A is found primarily in the intestine and in the regions of the brain that have serotonin, norepinephrine, dopamine and tyramine substrates; MAO-B is found primarily in platelets and in the regions of the brain that are rich in dopaminergic neurons.

The function of MAO inhibition was discovered in 1958 (Ott (1994:39). Drugs that suppress MAO-A in the brain have been used as anti-depressants since then. These drugs remain far longer in the system than plant-based sources. MAO inhibitors (MAOIs) inhibit the MAO enzyme, thereby allowing serotonin, norepinephrine and dopamine to accumulate in the synapse.[21] Mood changes induced by MAOIs can be aphrodisiac, and the plants considered below are sometimes used for that purpose in various cultures around the world.

14.2 *Plants containing monoamine oxidase inhibitor (MAOI)*

Around seventy plants are now known to contain monoamine oxidase inhibitors (MAOIs).[22] The list of plants known to contain chemicals that have this function augments nearly annually.

Probably the most commonly used plants that are used for their psychoactive and MAOI properties are Syrian rue (*Paganum harmala*) and *yagé* (*Banisteriopsis caapi* vine). Some plants containing small amounts of MAOI, such as nutmeg, liquorice and passion flower, are mainly used as a spice in cooking, as food, or in folk medicine. Others, such as caltrop, cowhage, yohimbe and black pepper, frequently feature in ancient and modern aphrodisiac formulas. Besides the plants considered below, other plants also have MAOI properties.

Syrian rue (Peganum harmala)

One of the MAOI plants that has a very long history of use is Syrian/mountain/wild rue (*Peganum harmala*).[23] This plant belongs to the *Zygophyllaceae* botanical family and is known

---

[20] Oxidases are enzymes that utilize molecular oxygen as acceptor, and convert phenolic substances to quinones.

[21] Culpepper (2013:3); Voogelbreinder (2009:22). See Snyder (1986:98–100) on how the enzyme MAO is responsible for the oxidative deamination of the neurotransmitters norepinephrine and serotonin.

[22] Ott (1994:73–75) lists sixty-seven plants in nineteen botanical families with MAO-inhibiting ß-carbolines.

[23] There is archaeological evidence from the Caucasus region of its use, probably as an intoxicant that was burnt and inhaled, which dates from the 5th millennium BCE (Sherratt 1995:30). In the 3rd millennium BCE, rue appears in a Mesopotamian cuneiform text as *šibbaratu*, in a section on poisoning (Scurlock 2014:636). Around 2,000 years ago, the Copts in Egypt used rue primarily as a medicine to treat skin diseases, worms and 'sick testicles' (Manniche 1998:145). Rue was well known to physicians in the classical Greco-Roman world. Pliny the Elder (23–79 CE) considered rue (*ruta*) to be among the chief medicinal plants, with numerous applications, and also as one of the main ingredients of antidotes to poisoning (*Natural History* 20.153). Dioscorides (*c.* 30–90 CE) records several, similar applications to those of Pliny, and notes that the wild or mountain varieties are referred to as *moly* by Cappadocians and others (*De Materia Medica* 3.51–54) (2000:423–428).

as *harmal/harmel* (in Arabic) or as *isphand/isfand/spand/sipand* more widely in the Middle-East and Asia. In Egypt it is known as *besata* ('Plant of Bes') and in Morocco as *mejnenna*, which means 'what makes your crazy/possessed' (Samorini 2019). It is probably the MAOI plant most widely used for its psychoactive, aphrodisiac and medicinal properties.

In Iran and neighbouring regions, the seeds of Syrian rue are still used for apotropaic purposes, and as an aphrodisiac—sometimes in the form of an extracted oil—by Turks, Moroccans, Tunisians and others (Flattery and Schwarz 1989:42–92; Samorini 2019). The shamans of Hunza, in the mountains of northern Pakistan, inhale the vapours of rue, which they call *supándur*, to 'call the spirits' during their trance (Samorini 2019). The inhalation of thick smoke from burning seeds is also mentioned in classical Persian poetry (Flattery and Schwarz 1989:47).

The angular, reddish brown seeds of rue are usually boiled before being crushed, producing a reddish extrusion, which is used as a traditional dye for cloth and carpets. The seeds contain almost equal quantities of the alkaloids harmine, harmaline, and tetrahydroharmine (THH), which are ß-carboline derivatives, all belonging to a class of compounds that act as monoamine oxidase inhibitors (MAOI). The effects of harmine, harmaline and tetrahydroharmine are similar but not identical. They induce an introverted, dream-like state, in which noise can be a disturbance. Naranjo (1973), who conducted pioneering experiments with harmine and hamaline in 1963, described the effects of these substances as 'oneirophrenic' (dream-inducing). Large doses can induce vomiting; this seems to be consistent with other MAOI plants, which can provoke purgation in high doses.

Yagé (Banisteriopsis caapi vine)

The *Banisteriopsis caapi* vine, which is in the *Malpighiacea* botanical family, grows principally in the Amazon region of South America, where it is usually known as *yagé*; it is also cultivated these days in several countries in South America and in other tropical regions, such as Hawaii. It is another rich source of MAOIs, though as a percentage of volume it is not quite as rich as in Syrian rue. Some varieties of the vine also contain traces of DMT and 5-MeO-DMT (see below).

The vine is used alone for its psychoactive effects by some groups, such as the Tukano in north-west Amazonia. It is macerated and left to stand in cold water, which is then drunk.[24] Other groups, such as the Matsigenka tribe of the Manu region of Peru, boil the vine until it attains a honey-like consistency; sometimes tobacco or other plants are added. From the 1960s onwards they began to add *Psychotria viridis* (Shepard (2018:75), usually known as *chacruna*, which contains *NN*-dimethyl tryptamine (DMT), which produces a psychedelic effect.[25]

---

[24] Beyer (2010:221); Millard (2018:83); Ott (1994:17); Ott (1996a:208); Stafford (1977:262); Voogelbreinder (2009:60).

[25] This concoction is commonly known as ayahuasca, a powerful psychedelic drink, which these days usually comprises not only a DMT plant (usually *Psychotria viridis*) and an MAOI plant (usually *Banisteriopsis caapi*), but also other additive plants. See Clark (2020; 2021) for further details. Torres (2018:39) observes that, "An investigation into the origins of *ayahuasca* reveals numerous beverages distributed throughout South America, each being distinct and varying according to plant availability, cultural predilections for ingestion and ritual requirements." Torres (p. 43–49) maintains that although some tryptamine-containing plants have been in use in South America since 2000 BCE, there is no evidence for use of the *Banisteriopsis caapi* vine for making ayahuasca before the initial contact with Europeans, *c.*1550–1650. However, Torres notes the use in South America of fermented drinks with an admixture of the DMT-containing seeds of *Anadenanthera peregrine* seeds, and in the

Fragrant ginger/galangal (Kaempferia galangal)

Another commonly used plant, which is rich in MAOIs, is galangal/aromatic ginger (*Kaempferia galangal*). It is native to South and South-East Asia and is used in Asian cuisine (Noro *et al.* 1983; Toro and Thomas 2007:46).

Nutmeg (Myristica fragrans)

The nutmeg tree (*Myristica fragrans*) is originally from southeast Asia, where it is commonly used as a spice in cooking and also as an aphrodisiac and a medicine (Rätsch and Müller-Ebelling (2013:466–467). The main chemical in nutmeg with MAOI properties is myristicin (Ott 1996:231). In India, nutmeg, known as *jāyphal* in Hindi, besides being a commonly used spice, is often found in multi-plant Āyurvedic aphrodisiac formulas. In this capacity, the nut itself is sometimes used separately from the hairy covering of the nut, known as *jāypatti* or *jāvitrī*, which is used for making the spice known as mace. Nutmeg was used in Europe in the Middle-Ages as an aphrodisiac. However, consuming large amounts of nutmeg can be poisonous.

Caltrop (Tribulus terrestris)

Tribulus terrestris, known as caltrop, puncture vine, cathead, and as *gokṣura* in Sanskrit, is another plant containing MAOIs (Ott 1994:75; Voogelbreinder 2009:333), in the form of harmine *etc.* It is commonly used in Indian Āyurvedic aphrodisiac formulas and also used by people hoping to increase athletic performance (Stuart 2021). It is a fruit-producing plant in the *Zygophyllaceae* botanical family that grows widely in the Mediterranean region, South Asia and South Africa.

Cowhage (Mucuna pruriens)

Mucuna pruriens, usually known as cowhage or Bengal velvet bean, contains not only MAOIs but also DMT and 5-MeO-DMT in trace amounts (Clark 2020:142; Ott 1994:74, 82); Trout 2002:122). This plant, in the *Fabaceae* botanical family, is known in Sanskrit as *kapicacchu* or *ātmaguptā*; it also features commonly in Āyurvedic aphrodisiac formulas. It is also used in Āyurveda to treat Parkinson's disease, as the plant also contains L-DOPA, a psychoactive chemical which is nowadays manufactured in pure form to treat the disease. In Indian folk medicine the plant is also used to treat snakebite.

Heart-leaved moonseed (Tinospora cordifolia)

Tinospora cordifolia (Sanskrit *guḍuci/madhuparṇi/soma/amṛta*, Hindi *giloy*, heart-leaved moonseed) contains jatorrhizine, berberine and palmatine, all of which, individually, act as MAOIs (both MAOI-A and MAOI-B. This plant has several Āyurvedic medical applications, including use for diabetes and as an anti-inflammatory (Kumar *et al.* 2020), and is also a common constituent of multi-plant aphrodisiac formulas.

---

mid and upper tributaries of the Orinoco, the simultaneous consumption of *Banisteriopsis caapi* and to the DMT-containing *yopo*, which could be seen as ayahuasca analogues.

14.3 *Other Commonly Used MAOI Plants*

Other, commonly used plants that have MAOI properties (see *Psychonautwiki* 2021), to varying degrees, include:

betel nut (*Areca catechu*) and yohimbe (*Pausinytalia johimbe*) [MAO-A inhibitors];

kava (*Piper methysticum*), olives (*Olea europaea*) [MAOI-B inhibitors];

black pepper (*Piper nigrum*), long pepper (*Piper longum*) cannabis (*Cannabis sativa*), cocoa (*Theobroma cacao*), coffee (*Coffea arabica, Coffea canephora*), golden root (*Rhodiola rosea*), tobacco [MOA-A and MAO-B].

Yohimbe was noted in Section 8.1 as a plant used for aphrodisiac purposes. It could also be seen that black pepper, long pepper and betel nut are among ingredients of both traditional (Section 11.4; 11.7) and modern Āyurvedic aphrodisiac formulas (12.1). I suspect that this is owing to their MAOI properties.

South Asian Plant Identifications

Caraka, vol. 4:287–317; Monier-Williams; *Pandanus Database of Indian Plants*; Sensarma; Sudarshan; Wisdom Library.

References

Abel, Ernest L. (1985). *Psychoactive Drugs and Sex.* New York/London: Plenum Press

Barillas, Christopher, V. J. (2022). 'Ketamine'. *Encyclopedia.com.* https://www.encyclopedia.com/science/applied-and-social-sciences magazines/ketamine

Beyer, Stephan V. (2010) [1999]. *Singing to the Plants: A Guide to Mestizo Shamanism in the Upper Amazon.* Albuquerque: University of New Mexico Press.

Bhambhvani, Hriday R., Alex M. Kasman, Genester Wilson-King, and Michael L. Eisenberg (2020). 'A Survey Exploring the Relationship Between Cannabis Use Characteristics and Sexual Function in Men'. *Sexual Medicine*, June 16, pp. 1–10. https://www.sciencedirect.com/science/article/pii/S2050116120300787

Block, Susan (2017). 'Cannabis and Sex'. In Mark J. Estren (ed.), *One Toke to God: The Entheogenic Spirituality of Cannabis*, pp. 143–148. Malibu CA: Cannabis Spiritual Centre.

Byers, Alton C. (2020). 'It's yarsa-picking time: A holy valley in eastern Nepal prepares for the yarsa gumba harvesting season'. *Nepal Times*, 20th March. https://www.nepalitimes.com/banner/its-yarsa-picking-time/

C., Hannah (2020). 'Why Television is Bad for Your Health'. *The Science Times*, 23rd July. https://www.sciencetimes.com/articles/26586/20200723/why-television-bad-health.htm

Canada Vigilance Program (2013). 'Archived – Health Canada warns consumers not to use Libidus or any unauthorized products promoted for erectile dysfunction'. Ottawa: Health Canada/Government of Canada (25th April). https://healthycanadians.gc.ca/recall-alert-rappel-avis/hc-sc/2008/13282a-eng.php

Caraka (trans. and ed. Priyavrat Sharma) (1994–1996). *Caraka-Samhitā (Agniveśa's treatise refined and annotated by Caraka and redacted by Dṛḍhabala, text with English translation)* (Jaikrishnadas Ayurveda Series 36), vols. 1–4. Varanasi/Delhi: Chaukambha Orientalia.

Chate, Vasudev, Praveen Bhirdi, Anand Katti, and Shreevatsa (2017). 'The Efficacy of Shatavari in Orgasm: A Comparative Clinical Study'. *Pijar (Paryeshana International Journal of Ayurvedic Research)*, vol. 1, issue 3, January–February, pp. 39–43.

Chauhan, Nagendra S., Vikas Sharma, and V. K. Dixit (2011). 'Effects of *Asteracantha longifolia* seeds on the sexual behaviour of male rats'. *Natural Product Research*, vol. 25, issue 15, pp. 1423–1431.

Cherdshewasart, W., and N. Nimsakul (2003). 'Clinical Trials of Butea superba, an alternative herbal treatment for erectile dysfunction'. *Asian Journal of Andrology*, vol. 5(3), September, pp. 243–246. https://pubmed.ncbi.nlm.nih.gov/12937809/

Choueke, Esmond (1998). '*Aphrodisiacs: A Guide to What Really Works.* Toronto: Citadel Press/Carol Publishing Group.

Clark, Matthew (2020) [2017]. *The Tawny One: Soma, Haoma and Ayahuasca.* London: Aeon Books.

——— (2021). *Botanical Ecstasies: Psychoactive Plant Formulas in India and*

Beyond. London: Psychedelic Press.
Corazza, Ornelia, Giovanni Martinotti, Rita Santacroce, Eleonora Chillemi, Massimo Di Giannantonio, Fabrizio Schifano, and Selim Cellek (2014). 'Sexual Enhancement Products for Sale Online: Raising Awareness of the Psychoactive Effects of Yohimbe, Maca, Horny Goat Weed and *Ginko biloba'*. *Biomedical Research International,* 15th June, 15841798. doi: 10.1155/2014/841798 https://www.ncbi.nlm.nih.gov/pmc/articles/PMC4082836/
Cordin, Eva (2021). 'Best Male Enhancement Pills of 2021: Top 8 Sex Supplements for Men'. *The Island Now,* 25th July. https://theislandnow.com/blog-112/best-male-enhancement-pills/
Culpepper, Larry (2013). ' Reducing the Burden of Difficult-to-Treat Major Depressive Disorders: Revisiting Monoamine Oxidase Inhibitor Therapy'. *The Primary Care Companion For CNS Disorders*, vol. 15(5), pp. 1–12. https://www.ncbi.nlm.nih.gov/pmc/articles/PMC3907330/ (accessed 3/08/2021)
Dalal, P. K., Adarsh Tripathi, and S. K. Gupta (2013). 'Vajikarana: Treatment of sexual disfunctions based on Indian concepts'. *Indian Journal of Psychiatry*, vol. 55 (Suppl. 2), January, pp. S273–S276.
Dalterio, Susan L. (1980). 'Perinatal or adult exposure to cannabinoids alters male reproductive functions in mice'. *Pharmacology Biochemistry and Behavior*, vol. 12, issue 1 (January), pp. 143–153.
Dalterio, S., A. Bartke, and S. Burstein (1977). 'Cannabinoids Inhibit Testosterone Secretion by Mouse Testes in Vitro'. *Science*, vol. 196, issue 4297 (June), pp. 1472–1473.
Dalterio, S., A. Bartke, C. Robertson, D. Watson, and S. Burstein (1978). 'Direct and pituitary-mediated effects of Δ_9-THC and cannabinol on the testis'. *Pharmacology Biochemistry and Behavior*, vol. 8, issue 6 (June), pp. 673–678.
Diocorides, Padanius (trans. Tess Anne Osbaldeston and R. P. A. Wood) (2000). *De Materia Medica.* Johannesburg: Ibidis.
Earlywine, Mitch (2002). *Understanding Marijuana: A New Look at the Scientific Evidence.* Oxford/New York: Oxford University Press.
Erowid (2021). 'Psychoactive Lotus/Lily'. https://www.erowid.org/plants/lotus/
Fadiman, James (2011). *The Psychedelic Explorer's Guide: Safe, Therapeutic, and Sacred Journeys.* Rochester, Vermont/Toronto: Park Street Press.
Fallah, Ali, Azadeh Mohmmad-Hasani, and Abasait Hosseinzadeh Colagar (2018). 'Zinc is an Essential Element for Male Fertility: A Review of Zn Roles in Men's Health, Germination, Sperm Quality, and Fertilization'. *Journal of Reproductive Fertility*, vol. 19(2), pp. 69–81. https://www.ncbi.nlm.nih.gov/pmc/articles/PMC6010824/ (accessed 22/12/2021)
Flattery, David Strophlet, and Martin Schwarz (1989). *Haoma and Harmaline: The Botanical Identity of the Indo-Iranian Hallucinogen "Soma" and its Legacy in Religion, Language and Middle Eastern Folklore* (Near Eastern Studies, vol. 21). Berkeley/Los Angeles/London: University of California Press.
Golas, Thaddeus (1978) [1971]. *The Lazy Man's Guide to Enlightenment.* Palo Alto, CA: Seed Centre
Grof, Stanislav (2019). The Way of the Psychonaut: Encyclopedia for Inner Journeys, vol. 2. Santa Cruz, CA: Multidisciplinary Association for Psychedelic Studies (MAPS).
Hatsis, Thomas (2014). 'Psychoactive Potions in Medieval Magic and Witchcraft'. In Robert Dickins (ed.), *Psypress*, vol. II, pp. 67–80. Falmouth: Psychedelic Press.
Houston Press (2021). '7 Best Male Enhancement Supplements for Better Performance

(2021 Reviews)'. https://www.houstonpress.com/storyhub/best-male-enhancement-supplements (accessed 22/12/2021).

Jay, Mike (2010). *High Society: Mind-Altering Drugs in History and Culture*. London: Thames and Hudson Ltd.

Karki, Sanjaya, and Johanna Schaffner (2015). 'Yarsagumba: The Quest for Himalayan Viagra'. *Adventure Medic*, 13th November (accessed 22/11/2021) https://www.theadventuremedic.com/features/yarsagumba-the-quest-for-himalayan-viagra/

Kolodny, Robert C., William H. Masters, Robert M. Kolodner, and Gelsen Toro (1974). 'Depression of Plasma Testosterone Levels after Chronic Intensive Marihuana Use'. *The New England Journal of Medicine*, vol. 290 (April), pp. 872–874.

Kotta, Sabna, Shahid H. Ansari, and Javid Ali (2013). 'Exploring scientifically proven herbal aphrodisiacs'. *Pharmacognosy Review*, vol. 7(13), pp. 1–10.

Kumar, Pradeep, Madhu Kamle, Dipendra K. Mahato, Himashree Bora, Bharti Sharma, Prasad Rasane, and Vivek K. Bajpai (2020). '*Tinospora cordifolia* (Giloy): Phytochemistry, Ethnopharmacology, Clinical Application and Conservation Strategies'. *Current Pharmaceutical Biotechnology*, vol. 21, issue 12. DOI: 10.2174/1389201021666200430114547 (accessed 22/11/2021) https://www.eurekaselect.com/181533/article

Larson, Gerald James (1979) [1969]. *Classical Sāṃkhya: An Interpretation of its History and Meaning.* Delhi/Varanasi/Patna: Motilal Banarsidass.

Lee, William H., and Lynne Lee (1994). *The Encylopedia of Concentrated Aphrodisiacs.* New York: Instant Improvement, Inc.

Lilly, John C. (1997) [1988]. *The Scientist: A Metaphysical Autobiography.* Berkeley CA: Ronin Publishing, Inc.

L. S. (2016) [2015]. *Sacred Journeys: Tripguide for Psychonauts.* Netherlands: Onderstroomboven Collectief.

Macmillan, Amanda (2017). 'Watching Too Much TV Is Bad for You, Even If You Also Exercise' *Time*, 14th November. https://time.com/5023129/tv-blood-clots/

Manniche, Lise (1990) [1998]. *An Ancient Egyptian Herbal.* London: British Museum Press.

Men's Journal (2021). '7 Best Male Enhancement Pills of 2021'. https://www.mensjournal.com/health-fitness/best-male-enhancement-pills/

Millard, Dale (2018). 'Broad Spectrum Roles of Harmine in *Ayahuasca*'. In Ghillean Prance *et al.* (eds.), *Ethnopharmacologic Search for Psychoactive Drugs (50th Anniversary Symposium, June 6–8 2017),* vol. 2, pp. 70–81. London/Santa Fe: Synergetic Press, Ltd.

Miller, Richard Alan (1985). *The Magical and Ritual Use of Aphrodisiacs.* New York: Destiny Books.

Monier-Williams, Monier (1994) [1899]. *Sanskrit-English Dictionary.* New Delhi: Munshiram Manoharlal Publishers Pvt. Ltd.

Müller-Ebeling, Claudia, Christian Rätsch, and Wolf-Dieter Storl (trans. Annabel Lee) (2003) [1998]. *Witchcraft Medicine: Healing Arts, Shamanic Practices, and Forbidden Plants*. Rochester, Vermont: Inner Traditions.

Naranjo, Claudio (1973). *The Healing Journey: New Approaches to Consciousness.* New York: Ballantine Books.

Nath, R. (2008) [2005]. *The Private Lives of the Mughals of India (126–1803 A.D.)*. New Delhi: Rupa & Co.

Noro, Tadataka, Toshio Miyase, Masanori Kuroynage, Akira Ueno, and Sego

Fukushima (1983) 'Monoamine Oxidase Inhibitor from the Rhizome of *Kaempferia galanga*'. *Chemical and Pharmaceutical Bulletin*, vol. 31, no. 8, pp. 2708–2711.

O'Shaughnessy, William Brooke (1843). 'On the Preparations of the Indian Hemp, or Gunjah, (*Cannabis Indica*): Their Effects on the Animal System in Health, and their Utility in the Treatment of Tetanus and other Convulsive Diseases'. *Provincial Medical Journal and Retrospect of the Medical Sciences*, no. 123 (Feb. 4th), pp. 363–369.

Osto, Douglas (2019). *Altered States: Buddhism and Psychedelic Spirituality in America.* New York/Chichester (UK). Columbia University Press.

Ott, Jonathan (1994). *Ayahuasca Analogues: Pangœn Entheogens.* Kennewick, WA: Natural Products Co.

——— (1996) [1993]. *Pharmacotheon: Entheogenic Drugs, their Plant Sources and History*, 2nd edn. Kennewick, WA: Natural Products Co.

Pandanus Database of Indian Plants (2021). http://iu.ff.cuni.cz/pandanus/database/. Prague: Seminar of Indian Studies, Institute of South and Central Asia, Faculty of Arts, Charles University.

Panyala, Nagender Reddy, Eladia M. Peña Méndez, and Josef Havel (2009). 'Gold and nano-gold in medicine: Overview, toxicology and perspectives. *Journal of Applied Biomedicine*, vol. 7(2), May, pp. 75–91.

Pathak, Panjak, B. R. Guru Prasad, N. Anjaneya Murthy, and S. N. Hegde (2011). 'The effect of *Emblica officinalis* diet on lifespan, sexual behaviour, and fitness characters in *Drosophila melanogaster*'. *Ayu*, vol. 32(2), April–June, pp. 279–284.

Pliny (trans. W. H. S. Jones) (1961). *Natural History*, Books 20–23, 24–27 (vols. VI and VII, Loeb Classical Library). Cambridge, Massachusetts/London: Harvard University Press/William Heinemann Ltd.

Plotkin, Mark J. (1993). *Tales of a Shaman's Apprentice: An Ethnobotanist Searches for New Medicines in the Amazonian Rain Forest.* Harmondsworth, Middlesex, UK: Penguin Books.

Plotkin, Mark J., Brian Hettler, and Wade Davis (2017). '*Viva* Schultes – A Retrospective'. In Ghillean Prance *et al.* (eds.), *Ethnopharmacologic Search for Psychoactive Drugs (50th Anniversary Symposium, June 6–8 2017),* vol. 2, pp. 95–119. London/Santa Fe: Synergetic Press Ltd.

Psychnautwiki (2021). 'MAOI'. https://psychonautwiki.org/wiki/MAOI (accessed 22/12/2021)

Rätsch, Christian, and Claudia Müller Ebeling (trans. Aida Sepic Williams) (2013) [2003]. *The Encyclopedia of Aphrodisiacs: Psychoactive Substances for Use in Sexual Practices.* Rochester, Vermont/Toronto: Park Street Press.

Rathva, Bhargavi, Bhavisha Patel, Jugruti Maurya, and Kinjal Bera (2017). 'Standardization and Evaluation of some Parameters of Adapotogenic Polyherbal Oral Dosage Form'. *International Journal of Pharmaceutical Sciences Review and Research*, vol. 42(1), January–February, pp. 1–7.

de Ropp, Robert S. (1957). *Drugs and the Mind.* London: The Scientific Book Club.

Roy, Sagor Chandra, Md. Mamum Sikder, Arjyabrata Saker, Md. Afaz Uddin, Neshat Masud, Md. Rakib Hasan, Nilay Saha, and N. S. K. Choudhuri (2017). 'Preclinical Anemia Panel Studies of "Makardhvaja" after Chronic Administration to Male Sprague-Dawley Rats'. *Biology and Medicine* (Aligarh), vol. 9, issue 2, pp. 1–4.

Samorini, Giorgio (trans. Tami Calliope) (2002). *Animals and Psychedelics: The Natural World and the Instinct to Alter Consciousness.* Rochester, Vermont: Park Street Press.

——— (2019). 'Peganum Harmala, the "Ayahuasca" of North Africa and Eurasia'. *Kahpi: The Ayahuasca Hub* (accessed 11/12/2021). https://kahpi.net/syrian-rue-peganum-harmala-ayahuasca/

Sastry, Vaidya Pammi Satyanarayana (2010). *Rāja Mārtaṇḍa of Bhoja: Text, Tanslation, English Translation with glossary ands explanatory notes on few technical words and few doubtful herbs* (Krishnadas Ayurveda Series 132). Varanasi: Chowkhamba Krishnadas Academy.

Schultes, Richard Evans, Albert Hofmann, and Christian Rätsch (2001) [1992]. *Plants of the Gods: Their Sacred, Healing, and Hallucinogenic Powers.* 2nd edn. Rochester, Vermont: Healing Arts Press.

Scurlock, JoAnn (2014). *Sourcebook for Ancient Mesopotamian Medicine.* Atlanta, Georgia: SBL Press.

Selden, Gary (1979). *Aphrodisia: A Guide to Sexual Foods, Herbs & Drugs.* New York: E. P Dutton.

Sensarma, P. (1992). 'Plant Names – Sanskrit and Latin'. *Ancient Science of Life*, vol. XII, nos. 1 & 2 (July–October), pp. 201–220.

Sharma, Rohit, and P. K. Prajapati (2017). 'Comparative Antimicrobial Screening of *Satva* (Sedimented Starchy Aqueous Extract) and *Ghana* (Solidified Aqueous Extract) of *Guduchi* (*Tinospora Cordifolia* (Willd.) Miers)'. *Innovare Journal of Ayurvedic Sciences*, vol. 5, issue 1, pp. 1–4.

Shepard, Glenn H. (2018). 'Spirit Bodies, Plant Teachers and Messenger Molecules in Amazonian Shamanism'. In Ghillean Prance *et al.* (eds.), *Ethnopharmacologic Search for Psychoactive Drugs (50th Anniversary Symposium, June 6–8 2017),* vol. 2, pp. 70–81. London/Santa Fe: Synergetic Press, Ltd.

Sherratt, Andrew (1995). 'Alcohol and its Alternatives: Symbol and Substance in Pre-Industrial Cultures'. In Jordan Goodman, Paul E. Lovejoy, and Andrew Sherratt (eds.), *Consuming Habits: Drugs in History and Anthropology*, pp. 11–46. London/New York: Routledge.

Shilpa, Patil, and A. K. Chaudhari (2013). 'Pharmaceutical Standardization of Guduchi Satva'. *International Journal of Pharmaceutical & Biological Archives*, vol. 4 (1), pp. 109–113.

Shulgin, Alexander (2021). *The Nature of Drugs: History, Pharmacology, and Social Impact.* Berkeley/Santa Fe/London: Transform Press/Synergetic Press.

Shulgin, Alexander, and Ann Shulgin (1997). *TiKHAL: The Continuation.* Berkeley: Transform Press.

——— (2000) [1989]. *PiKHAL: A Chemical Love Story.* Berkeley: Transform Press.

Siegel, Ronald K (2005) [1989]. *Intoxication: The Universal Drive for Mind-Altering Substances.* Rochester, Vermont: Park Street Press.

Singh, Rita (2006). *Psychoactive Medicinal Plants: Hallucinogens and Narcotic Drugs.* New Delhi: Global Vision Publishing House.

Snyder, Solomon H. (1986). *Drugs and the Brain.* New York: Scientific American Library.

Sood, S. K., Sarila Rana, and T. N. Lakhanpal (2005). *Ethnic Aphrodisiac Plants.* Jodhpur: Scientific Publishers (India).

Stafford, Peter (1977). *Psychedelics Encyclopedia.* Berkeley, CA: And/Or Press.

Stolaroff, Myron J. (1994). *Thanatos to Eros: Thirty-Five Years of Psychedelic Exploration.* Berlin: VWB.

Stuart, Annie (2021). 'Tribulus Terrestris'. *WebMD*, 29th July. https://www.webmd.com/vitamins-and-supplements/tribulus-terrestris-uses-and-risks

Sudarshan, S. R. (2005). *Encyclopaedia of Indian Medicine*, vol. 4: *Materia Medica – Herbal Drugs.* Mumbai: Ramdas Bhatkal/Popular Prakashan Pvt. Ltd.

Suśruta (trans. and ed. Kaviraj Kunja Lal Bhishagratna) (1911). *The Sushruta Samhita (with a full and comprehensive introduction, additional texts, different readings, notes, comparative views, index, glossary and plates)*, vols. 1–3. Calcutta: Kaviraj Kunja Lal Bhishagratna.

Taberner, Peter V. (1985). *Aphrodisiacs: The Science and the Myth*. Philadelphia: University of Pennsylvania Press.

Tart, Charles T. (1971). *On Being Stoned: A Psychological Study of Marijuana Intoxication*. Palo Alto, California: Science and Behavior Books.

The Hans India (2017). 'Alstonia Scholaris plant causing health hazards'. 29th November. https://www.thehansindia.com/posts/index/Andhra-Pradesh/2017-11-29/Alstonia-Scholaris-plant-causing-health-hazards/341901?infinitescroll=1

Toro, Gianluca, and Benjamin Thomas (2007). *Drugs of the Dreaming, Oneirogens:* Salvia divinorum *and other Dream-Enhancing Plants*. Rochester, Vermont: Park Street Press.

Torres, Constantino Manuel (2018). 'From Beer to Tobacco: A Probable Prehistory of *Ayahuasca* and *Yagé*'. In Ghillean Prance *et al.* (eds.), *Ethnopharmacologic Search for Psychoactive Drugs (50th Anniversary Symposium, June 6–8 2017)*, vol. 2, pp. 36–69. London/Santa Fe: Synergetic Press, Ltd.

Trout, Keeper of the, *et al.* (2002) [1993–2001]. *Trout's Notes on Some Simple Tryptamines: A brief overview and resource compendium* (Trout's Notes #FS-X7, Version 4-2002). Mydriatic Productions/Better Days Publishing.

Turner, D. M. (1994). *The Essential Psychedelic Guide*. San Francisco: Panther Press.

Vāgbhaṭa (trans. Kanjiv Lochan) (2017). *Aṣṭāṅga Hṛdayam of Vāgbhaṭa: Sutra Sthana* (The Mohandas Indological Series 12). New Delhi: Chaukhambha Publications.

Voogelbreinder, Snu (2009). *Garden of Eden: The Shamanic Use of Psychoactive Flora and Fauna, and the Study of Consciousness*. (Self-published).

Whan, Lynne B., Mhairi C. L. West, Neil McClure, and Sheena E. M. Lewis (2006). 'Effects of delta-9-tetrahydrocannabinol, the primary psychoactive cannabinoid in marijuana, on human sperm in vitro'. *Fertility and Sterility*, vol. 85, issue 3 (March), pp. 653–660.

Wisdom Library (2021). https://www.wisdomlib.org

Wise, Jeff (2016). 'The Obscure Legal Drug That Fuels John McAfee'. *Intelligencer*, 30th September. https://nymag.com/intelligencer/2016/09/the-obscure-legal-drug-that-fuels-john-mcafee.html

Wujasyk, Dagmar (2015). 'On Perfecting the Body. Rāsayana in Sanskrit Medical Literature'. In *AION: Annali dell'Università degli Studi di Naploli "L'Orientale". Elisir Mercuriale e Immortalità. Capitoli per una Storia dell'Alchimia nell'Antica Eurasia. A cura di Giacomella Orofino, Amneris Roselli e Antonella Sannino* XXXVII, 2015, pp. 55–57.

——— (2017). 'Acts of Improvement: On the Use of Tonics and Elixirs in Sanskrit Medical and Alchemical Literature'. *History of Science in South Asia (Special Issue: Transmutations: Rejuvenation, Longevity, and Immortality in South Asia and Inner Asia)*, eds. Dagmar Wujastyk, Suzanne Newcombe, and Christèle Barois, pp. 1–36.

Wujastyk, Dominik (1998). *The Roots of Āyurveda: Selections from Sanskrit Medical Writings*. New Delhi/London/New York: Penguin Books.

www.ingramcontent.com/pod-product-compliance
Lightning Source LLC
Chambersburg PA
CBHW081057170526
45166CB00006B/2098